Color and Light in Nature

This beautiful and informative guide provides clear explanations to all naturally occurring optical phenomena seen with the naked eye, including shadows, halos, water optics, eclipses and mirages. Separating myth from reality, it outlines the basic principles involved, and supports them with many figures and references, and a wealth of rare and spectacular photographs. There are many helpful hints on how best to observe and photograph the phenomena. For this second edition, the authors have added over 50 new images and provided extra material on experiments you can try yourself.

DAVID LYNCH is an astronomer and atmospheric physicist, specializing in infrared studies of star-formation regions, comets, supernovae and cirrus clouds. After receiving his PhD in astronomy from the University of Texas at Austin, he was a Visiting Associate in Physics at the California Institute of Technology and an Assistant Astronomer at the University of California at Berkeley. He has worked at the Hughes Research Laboratories and the Aerospace Corporation, and operates Thule Scientific, a private research institute. He has organized and chaired many meetings on atmospheric science and been Principal Investigator on a variety of NASA, NOAA, NSF, and DoD programs.

BILL LIVINGSTON has been an astronomer at the Kitt Peak Observatory in southern Arizona for his entire professional life (since January 1959). Originally he helped design and build instruments and telescopes. Later he became a solar observer and has spent a great deal of time on Kitt Peak, a fact which accounts for many of the pictures in this book. He participated in many solar eclipse expeditions which took him to Alaska, the South Pacific, Africa, Indonesia, India, and recently Turkey. He has traveled widely in Russia and China, but he hastens to add that his best sightings of atmospheric phenomena were from his backyard in Tucson.

Second Edition

Color and Light in Nature

David K. Lynch and William Livingston

PUBLISHED BY THE PRESS SYNDICATE OF THE UNIVERSITY OF CAMBRIDGE
The Pitt Building, Trumpington Street, Cambridge, United Kingdom

CAMBRIDGE UNIVERSITY PRESS
The Edinburgh Building, Cambridge CB2 2RU, UK
40 West 20th Street, New York, NY 10011-4211, USA
10 Stamford Road, Oakleigh, VIC 3166, Australia
Ruiz de Alarcón 13, 28014 Madrid, Spain
Dock House, The Waterfront, Cape Town 8001, South Africa

http://www.cambridge.org

© Cambridge University Press 1995, 2001

First published 1995
Second edition 2001

Printed in the United Kingdom at the University Press, Cambridge

Typeface Swift Regular 9.5/14pt.   *System* QuarkXPress®   [UPH]

A catalogue record for this book is available from the British Library

Library of Congress Cataloguing in Publication data

Lynch, David K., 1946–
    Color and light in nature / David K. Lynch, William Livingston. – 2nd ed.
        p.   cm.
    Includes bibliographical references and index.
    ISBN 0 521 77284 2 – ISBN 0 521 77504 3 (pb)
    1. Light. 2. Optics. 3. Color. 4. Astronomy. I. Livingston, W. C. (William
Charles), 1927– II. Title.

QC335.2.L96 2001
535–dc21   00-064230

ISBN 0 521 77284 2 hardback
ISBN 0 521 77504 3 paperback

The beholding of the light is itself a more excellent and a fairer thing than all the uses of it.

Francis Bacon, Novum Organum, Book 1

Contents

Preface to the second edition xi
Preface to the first edition xiii

1 Shadows 1
1.1 What is a shadow? 1
1.2 Brightness of shadows 4
1.3 Colored shadows 4
1.4 Contrail shadows 4
1.5 Opposition effect 6
1.6 Shadows and perspective 7
1.7 Spectre of the Brocken 11
1.8 Mountain shadows 12
1.9 Rays 16
1.10 Other shadow phenomena 19
      References 19

2 Clear air 21
2.1 Sunlight and atmospheric absorption 21
2.2 The clear blue sky 22
2.3 The sky near the horizon 24
2.4 Sky polarization 26
2.5 Airlight 29
2.6 Color and brightness of the low sun 31
2.7 The aureole 32
2.8 Bishop's ring 33
TWILIGHT 33
2.9 Twilight 33
2.10 Guide to twilight 34
2.11 Twilight arch 36
2.12 Earth shadow and the antitwilight arch 38
2.13 Alpenglow 41
2.14 Purple light 43
2.15 'It's darkest just before dawn' 43
2.16 Twilight and volcanic activity 44

NORMAL ATMOSPHERIC REFRACTION 46

2.17 Atmospheric refraction 46

2.18 Horizons 47

2.19 Flattening of the low sun and moon 47

2.20 Green flash 49

2.21 Twinkling 53

UNUSUAL ATMOSPHERIC REFRACTION 54

2.22 Distortions of the low sun 54

2.23 Mirages 55

2.24 Inferior mirage 55

2.25 Superior mirage 58

2.26 Lateral mirage 62

2.27 Airglow 63

2.28 Aurora borealis (northern lights) 63

References 69

3   Water and light 71

3.1   Light from water 71

3.2   Color of pure water 72

3.3   Color from suspended particles 73

3.4   Red tide and phosphorescent seas 75

3.5   Myth of a mirror image 77

3.6   Refraction through the air–water interface 78

3.7   Optical manhole 79

3.8   Polarization and fish-locating goggles 80

3.9   Visibility of waves and the horizon 80

3.10 Wave streaming near the horizon 81

3.11 Glitter 83

3.12 Skylight reflected from wavy water 87

3.13 Moon circles 88

3.14 Skypools and landpools 89

3.15 Shift of reflected skylight towards the horizon 92

3.16 The caustic network 93

3.17 Cat's paws 96

3.18 Slicks and oil-on-troubled-water 97

3.19 Why is foam white? 99

3.20 The wet spot 100

3.21 Shadows on water 101

3.22 Aureole effect 102

3.23 Other reflections on water 103

References 108

4   Water drops 109

RAINBOWS AND THEIR KIN 109

4.1   Sparkling dewdrops 109

4.2   Observations of rainbows 109

4.3   The primary rainbow 116

4.4   The secondary rainbow 119

4.5   Supernumerary bows 119

4.6   Tertiary bows 122

4.7   Alexander's dark band 122

4.8   Polarization of rainbows 123

4.9   Reflection rainbows 124

4.10 Fogbows 125

4.11 Cloud contrast bows 127

FORWARD AND BACKSCATTER PHENOMENA 128

4.12 Heiligenschein 128

4.13 Coronae 129

4.14 Irisation 133

4.15 The glory 135

4.16 Water drop optical effects 137

CLOUDY SKIES 139

4.17 What is a cloud? 139

4.18 Why are clouds white? 140

4.19 Why are some clouds dark? 140

4.20 Colors of clouds 143

4.21 Cloud blocking 144

4.22 Blinks 144

4.23 Does every cloud have a silver lining? 147

4.24 Walking in a fog 148

4.25 Visibility of the sun through a cloud 149

4.26 Once in a blue moon 149

4.27 Haze, smog, and smoke 150

4.28 Contrails and distrails 150

ELECTRICITY 151

4.29 Lightning 151

4.30 Sprites and jets 153

References 157

5   Ice and halos   159

5.1   Glints and sparkles   159

5.2   Ice and its optical properties   160

5.3   Color in snow banks and glaciers   161

5.4   Introduction to halos   163

5.5   The 22° halo   166

5.6   The 46° halo   169

5.7   Circular halos of unusual radii   170

5.8   Parhelia (sundogs)   171

5.9   Paranthelia and paranthelic arcs   175

5.10   Circumzenithal arc and circumhorizontal arc   176

5.11   Pillars   178

5.12   Parhelic circle   180

5.13   Tangent arcs of the 22° halo (circumscribed
             halo of 22°)   180

5.14   Subsuns and Bottlinger's rings   183

5.15   Halos below the horizon   184

RARE AND UNUSUAL HALOS   184

5.16   Parry arcs   184

5.17   Heliac arcs   185

5.18   Anthelion   185

5.19   Anthelic pillar   185

5.20   Tangent arcs of the 46° halo   185

5.21   Lateral arcs of the 22° halo (Lowitz arcs)   186

5.22   Lateral arcs of the 46° halo   186

5.23   Anthelic arcs   186

5.24   Multiple halo displays   187

5.25   Multiple scattering effects   187

5.26   Diffraction halos   187

5.27   Elliptical halos   187

5.28   Halo catalog   188

THE HIGHEST CLOUDS   192

5.29   Noctilucent clouds ('night-shining' clouds)   192

5.30   Nacreous clouds   193

             References   193

6   Naked-eye astronomy   197

DAYTIME   197

6.1   The sun   197

6.2   Sunspots   198

6.3   Solar eclipses   199

6.4   Total solar eclipse phenomena   200

6.5   Eclipse danger to the eye   204

6.6   Unusual eclipse phenomena   204

6.7   Annular and partial eclipses   205

6.8   Future solar eclipses   206

NIGHTTIME   208

6.9   Artificial satellites   208

6.10   The moon   211

6.11   Phases of the moon   212

6.12   Earthshine   214

6.13   Lunar eclipses   214

6.14   The planets   216

6.15   Meteors   217

6.16   Grazing meteors   218

6.17   Comets   218

6.18   Zodiacal light and gegenschein   220

6.19   Stars and the stellar magnitude scale   221

6.20   Daytime visibility of stars   223

6.21   Constellations and star names   223

6.22   Double stars, variable stars, novae, and supernovae   223

6.23   Star clusters and nebulae   223

6.24   The Milky Way   224

6.25   Galaxies   224

6.26   Light of the night sky   224

6.27   Urban glows   224

6.28   Starlight   226

6.29   Olbers' paradox   226

             References   227

7   Observing   229

HUMAN VISION   230

7.1   The eye   230

7.2   Photopic (cone) vision   230

7.3   Scotopic (rod) vision   232

7.4   Angular resolution of the eye   233

7.5   Time constants of the eye   233

7.6   Color vision   233

SUBTLETIES OF VISION 234

7.7   Mach bands 234

7.8   Irradiation 234

7.9   After-images 234

7.10 Illusions 235

7.11 Haidinger's brush 236

7.12 Floaters 237

OBSERVING TOOLS 238

7.13 Cameras 238

7.14 Polarizers 238

OBSERVING TECHNIQUE 239

7.15 Measurement of angles 239

7.16 Universal Time 240

7.17 Out-of-focus viewing 240

7.18 Observations from an airplane 244

      References 246

8   Exotic Clouds 247

      References 258

Glossary 259

Index 273

Preface to the second edition

This second edition is an opportunity to greatly enlarge the pictorial content of the original. We appreciate contributions by Peter Athans, Detrick Branston, Bart Cardon, Alan Clark, Arthur Clarke, Jan Curtis, Rayleigh Drake, Bruce Gillespie, David and Gary Gutierrez, Joe Hickox, Charles Hunter, Dick Hutchinson, Gary Ladd, Ken Langford, Peter Livingston, Bruce McKibben, Marko Pekkola, Jean Rösch, Barbara Schaefer, Pierre Turon, Dale Vrabec, and Andy Young. Robert O'Shea, Jack Harvey, Bernard Soffer, Kurt Nassau, and Andy Young made useful critical remarks.

Our hope is that Minnaert's spirit of looking and questioning is preserved. May the reader discover new phenomena that we haven't thought of!

W CL wcl@noao.edu
D KL thule@earthlink.net

Preface to the first edition

This book is about seeing the world with the naked eye. Everyone likes a beautiful sunset. But if you are an astronomer (like we are), you look at the sky *very* carefully. And sometimes worry. Will it be clear tonight? What kind of clouds are those and what do they mean? Will those clouds on the horizon move in and shut us down? Will the wind pick up and force us to close the telescope? About all you can do is watch but after a lifetime of scrutinizing the sky, you start seeing things. This book is about those things.

For me (Bill) it started on Mount Wilson Observatory, near Los Angeles. I climbed atop one of its pinnacles and saw my shadow thrown on a bellowing fog bank. I was startled by its monstrous dimensions, and as I waved my arms about (that's something you learn to do in fog) the shadow copied my every motion – in three dimensions! This was my first Spectre of the Brocken.

For me (this is Dave talking now) it began one cloudy night in west Texas while I was grousing about the McDonald Observatory looking for something to read. A color photograph of a light pillar on a copy of *American Scientist* caught my eye. I turned to the article and found Bob Greenler's now famous light pillar article. I remember reading the article over and over, hardly able to sit still long enough to finish it. So that's what I saw that winter morning fifteen years earlier! I was hooked.

By the time we met in the early 1970s, our dog-earred copies of Minnaert's *The Nature of Light and Colour in the Open Air* were faithful companions. In 1975 we were riding up to Kitt Peak with several other astronomers and we were discussing the preparations for the observing run on the McMath solar telescope. During a lull in the conversation I mentioned that Dale Vrabec had shown me some interesting light circles made when the moon reflects off the ocean (moon circles). Suddenly Bill's eyes lit up. Astronomy was forgotten and he wanted to know everything about what we saw. That's how this book began.

We became acquainted with Minnaert and found a rich and deeply curious man. He was something of a bookworm and his observatory in Utrecht, Holland contains a treasure trove of mete-orological material. His book, prepared during the scientific pause forced by World War II (during which he was jailed by the Nazis), uniquely brought together his observations. Our title, *Color and Light in Nature* is a clumsy but heart-felt parsing of his title. But we still like his best.

This book was inspired by Minnaert but enrichment came from many sources. We recall: Alistair Fraser's print exhibit at Boulder's NCAR building; a visit with Ron Tricker on the Isle of Wight in 1978; the Optical Society of America's topical meetings on 'Meterological Optics' and 'Color and Light in the Open Air' (titles borrowed from books by Tricker and Minnaert, respectively); UCLA Extension who allows one of us (DKL) to teach a course on this subject. We have been honored to work with some wonderful people and we would like to thank Craig Bohren, Alistair Fraser, Douglas Gough, Bob Greenler, Freeman Hall, Mark Hanna, Glen Pickens, Jarus Quinn, Ken Sassen and Ron Tricker. These good folks made invaluable contributions to the book. If asked about their part, they would probably shrug it off as simply something that friends do. But we know that this book would not be here without them.

Now it's up to you. Go outside, read the book and enjoy the light.

Topanga, California   DKL
Tucson, Arizona   WCL

1 Shadows

A number of contrails already crisscrossed the sky when we spied an eastbound jet laying out yet another. The winds aloft carried this fresh contrail across the solar disk. At the instant of transit, the earth and sky were stabbed by a dark, sword-like shaft. In less than a second, it was over and the shadow swept silently, invisibly up the canyon. Like an inverse lightning bolt, the airborne shadow caught us by surprise and left us wondering.

Adapted from the authors' observing books

1.1 What is a shadow?

We think of shadows as dark areas on the ground. But they are also volumes of space that are shielded from light (Figures 1.1A, 1.1B). This darkened space is normally invisible but when enough particles are present to scatter light from it, the shadow's three-dimensional character is revealed.

There are two kinds of shadows: those from point sources which are simple and rather rare, and those from extended sources like the sun which are common but more complex. If the light source is a point source such as a star, the shadows are dark and have sharp edges (Figure 1.1C(a)). This shadow is called an umbra. Occasionally the glint from a chrome automobile bumper or a distant windshield approximates a point source and produces sharp-edged shadows.

When the light source is extended like the solar disk, which is about 1/2° across, an object casts a two-part shadow, an umbra and a penumbra (Figure 1.1C(b)(c)). The umbra receives no direct sunlight, and therefore an observer in the umbra cannot see the sun. The penumbra, which is partially illuminated by the sun, is visible as a gradation of light around the umbra reaching outward toward fully illuminated space. From within the penumbra he sees a partially eclipsed sun. In everyday experience, 'shadow' and 'umbra' are often synonymous, and the fuzzy-edged penumbra goes unnoticed except to the careful observer.

Cast a shadow with your hand onto a piece of paper. See how the sharpness of the shadow's edge diminishes as you raise your hand above the surface. That fuzzy outer edge of the shadow is the penumbra. Its width divided by its distance from your hand is always a constant fraction of about 1/112. This is the angular diameter of the sun measured in radians.

Umbrae may or may not extend without limit into space. If the light source is physically larger than the shadower, the umbra tapers to a point behind the object. This is what happens with the

Fig. 1.1A Crepuscular rays fan out from clouds when viewed toward the sun. (Tucson, Arizona)

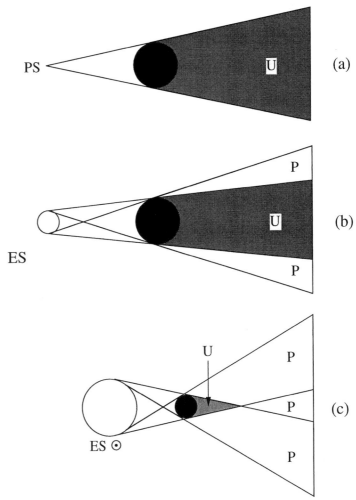

Fig. 1.1C Geometry of shadows. (a) A point source of light PS casts only an umbra U which is totally hidden from the point source. Except for a minor contribution from diffraction, point sources cast no penumbrae. (b) An extended source ES, like the sun, casts both an umbra U and penumbra P. (c) When the light source, like the sun (☉), is physically larger than the opaque object, the umbra has a finite length. When it is smaller, the umbra extends without limit behind the opaque object.

Fig. 1.1B Rays in Ponderosa Pines made visible by campfire smoke. The rays appear nearly parallel because our line-of-sight is perpendicular to the direction of the sun. Compare these rays with those in Figure 1.1A. (Yosemite, California)

sun and earth. The earth's umbra is totally enclosed by its penumbra. The apex of the umbral cone is 112 earth diameters behind the earth. As shown in Figure 1.1C(c), the penumbra is an ever-expanding region of space that surrounds the umbra. Beyond the tapered point of the umbra, the observer in the shadow (here a penumbra) will see the shadowing body silhouetted against the solar disk.

If there are multiple sources of light, as in a house at night, shadows become indistinct because the shadow from one object is illuminated by light from another source. The larger the size of the source, the smaller the umbra. On an overcast day, the whole sky is the light source and shadows are soft and subdued.

Believe it or not, a transparent object can cast a shadow. Look at the shadow from your eyeglasses or a drinking glass. This shadow is not due to blockage of light as with an opaque object, but rather to refraction. Refraction redirects the light and prevents it from falling where it would if it was traveling in a straight line.

1.2 Brightness of shadows

Shadows on earth are never totally dark. If they were, we could not see anything in them. When the astronauts visited the airless moon, they found the shadows to be very dark indeed. With no sky-light, there was little to illuminate the shadows. Were it not for earthshine, the shadows would have been completely dark. On earth, light reaching the umbra comes mostly from the sky. It bathes the shadow with blue light from nearly every direction.

1.3 Colored shadows

A shadow's color is determined by three conditions: (1) the color of the light shining into it; (2) the intrinsic color of the object on which the shadow is cast; and (3) a psychological factor called chromatic adaptation (Section 7.6). The simplest case is a shadow on snow. Snow is white because it scatters all colors equally. And shadows on snow reflect the light falling in them. On a clear day, shadows are tinted by blue skylight. In the presence of a low sun, the blue shadows stand out strongly against the yellow-tinged sur-roundings (Figure 1.3). On overcast or hazy days, bluish coloration is muted because the sky is much whiter.

While the blue sky adds a bluish tint to all daytime shadows, the color perceived by the eye depends in a complicated way on the brightness and color of the object and skylight (both of which con-

Fig. 1.3 Blue shadows of a person and a tree cast at low sun onto a white storage tank. This coloration is due to blue skylight which illuminates the shadow. (Kitt Peak, Arizona)

tain all colors). Only rarely does the color of a shadow differ markedly from the object on which it falls.

Sometimes, however, shadows do appear to have colors other than blue. Goethe has described green shadows in some detail[1]. The authors have also seen decidedly green shadows in the mountains at sunset. On subsequent evenings, at the same place, under what we thought were identical lighting conditions, the shadows were not green at all. Perhaps this is an example of chromatic adaptation (Section 7.6).

1.4 Contrail shadows

Jets often precipitate contrails (contraction for CONdensation TRAILS). Contrail shadows can be cast on the ground, thin lower

Fig. 1.4A Contrail shadow projected onto low level stratocumulus as seen from an aircraft. A glory (Section 4.15) is also visible at the antisolar point.

Fig. 1.4B Faint contrail shadow cast on hazy air. (Kitt Peak, Arizona)

clouds or a haze (Figures 1.4A, 1.4B). Airplane passengers can see their plane's contrail as a dark, narrow line trailing the shadow of the aircraft at the antisolar point.

A contrail shadow can be startling when observed from a point within it. In three dimensions, the contrail's shadow is like a curtain. It lies in a plane defined by the contrail and the sun. The reduction or absence of direct sunlight in the shadow decreases the amount of light scattered by particles within the shadow. From outside the shadow, it usually cannot be seen because its contrast is too low. This is because the line-of-sight through the shadow is only a few meters and continues for many kilometers through sunlit air so that the slight amount of lost light is negligible. But from within the narrow plane of the shadow, the line-of-sight can be enormous, in some directions many kilometers. Suddenly the shadow springs into view (i.e. its contrast increases dramatically) and it appears etched darkly across the sky.

Contrail shadows are seldom seen in the air because they are so narrow. A new contrail will be only 100 meters or so wide at its height of origin. Assuming an elevation of 8000 meters, the width of the umbra is only about 20 meters on the ground owing to its taper. Winds will push the shadow along at typically 20 meters per second. This means that the observer will only be in the umbra for less than a second. This is far too short a time to catch our attention unless it is expected.

A similar shadow can be seen while driving through the shadows of power lines and poles on a hazy or foggy day. As we pass through the plane of the power line's shadow, the dark umbra sweeps by.

1.5 Opposition effect

Have you ever been flying on a sunny day and seen a fuzzy bright spot accompanying you on the ground at the antisolar point? This is the opposition effect. It is a few degrees across and appears as a slight localized brightening of the terrain (Figure 1.5A). The opposition effect is present over all types of landscape but is especially noticeable in uniform vegetation. It would be harder to see if it wasn't moving. Only over water is the opposition effect absent.

Astronomers first noticed the opposition effect as a brightening of the moon and Mars near opposition. Opposition means 'opposite the sun as viewed from the earth'. Something is at opposition when it is at or near the antisolar point. The moon is substantially brighter around opposition (full moon) than can be accounted for by simply the increased area illuminated.

The opposition effect can have several causes, the most common being shadow hiding[2,3]. Shadow hiding creates a bright spot in the following way. The light reflected to our eye from any part of the landscape consists of both sunlit and shaded portions. On average the landscape is darker than the sunlit parts alone, and brighter than the shadows. At the antisolar point the observer's line-of-sight is coincident with the sun's rays and therefore shadows cannot be seen. In the absence of shadows the landscape's brightness is higher than average, creating a bright spot (Figure 1.5B). Although there is no vegetation on the moon or Mars, shadows produced by their rough surfaces are enough to result in significant brightening at the antisolar point.

On the earth, plants are plentiful sources of tiny shadows and the opposition effect is most evident over forests and prairies. A somewhat less noticeable opposition effect also occurs over barren ground, provided there is enough surface texture to cause shadows.

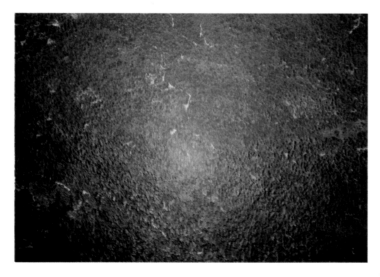

Fig. 1.5A The bright area in the middle of this forest viewed from an aircraft represents the opposition effect. The increase of brightness is due to the reduced visibility of the shadows near the antisolar point. There is also a slight color change because there are fewer shadows here. This makes the central region yellower compared to shadowed areas which are mainly illuminated by the blue sky.

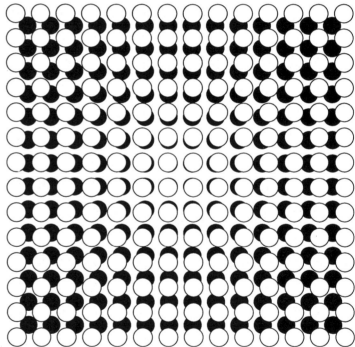

Fig. 1.5B Shadow hiding causes the opposition effect. All objects like branches, leaves, and rocks (represented by open circles) cast shadows (dark circles). When the observer looks towards the antisolar point, the line-of-sight parallels the sun's rays and the objects' shadows are hidden from view. This results in a brightness enhancement (center of picture).

The color of the bright spot differs from the average landscape scene because it contains none of the blue component from reflected skylight that is found in shadows. Thus opposition effects tend to be slightly yellower (i.e. less blue) in tone than their immediate surroundings; dark green forests appear light green in the bright spot, light green grass goes to yellow-green.

The shape of the opposition effect depends on the geometry of the shadowers. Above forested and bushy areas it is nearly circular. Over grass or cultivated fields where tall, slim plants grow parallel to one another and cast thin shadows, the opposition effect is noticeably oval with a vertical elongation.

While shadow hiding causes the opposition effect, at least three other optical mechanisms can produce brightening at the anti-solar point (Figure 1.5C). These effects involve the preferential scattering of light in the backward direction. They may involve backscattering in which light is scattered generally in the backward direction, or retro-reflection in which light is scattered precisely in the backwards direction.

Bare rocks can retro-reflect. Tiny crystal intrusions having rounded, weathered outer surfaces act like lenses that focus light. If there is a scatterer near the focus, it sends light in all directions, much of which falls on the lens which sends it back towards the sun[4,5]. Other rocks contain crystals that form corner cubes and retro-reflect incoming sunlight. These enhanced returns cannot be seen when we stand close to rocks because the angular size of our own shadow exceeds the size of the opposition effect; observations from aircraft are required. There is also a surprising variety of transparent spheres in nature (sap from pine trees, certain kinds of ice plants, etc.) that return light just as water drops do.

There is one more factor in the opposition effect: coherent backscatter[6]. Coherent backscatter is a newly-recognized interference-like component to diffuse scattering from granular surfaces like sand. Unlike in clouds where the scatterers are far apart, the particles are so close together that scattered waves from nearby particles interfere with one another. Coherent backscatter is strongest in the 180° direction and falls off quickly away from it.

Water drops on a surface produce a powerful and different kind of opposition effect called the heiligenschein (Section 4.12).

1.6 Shadows and perspective

When the sun is low our own shadow appears distorted (Figure 1.6A). We see ourselves as if on stilts; our legs stretch and turn spindly, our torsos shrink. Worst of all, we become pinheads (Figures 1.6B, 1.6C).

Distortion of the human shadow from a low sun is a matter of perspective. The actual shape of our shadow on the ground, say as seen from directly above, would be a stretched-out version of our true profile. When viewed from the level of our eye (it's hard to do

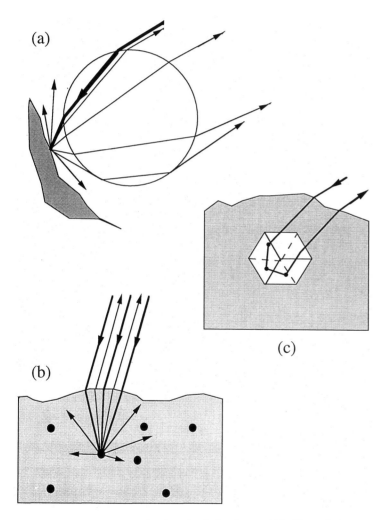

Fig. 1.5C Backscattering mechanisms in the landscape: (a) heiligenschein; (b) a scatterer lying at the focus of a rounded lens-like pebble in a rock; and (c) a crystal in the shape of a corner cube.

Fig. 1.6A The distorted shadows of people at low sun on a beach. Notice that most of the shadow comes from the lower parts of the body and that the shadows are generally triangular in outline as is the case for the mountain shadow seen in Figure 1.8A. (Puerto Peñasco, Mexico)

Fig. 1.6B (LEFT) Distorted shadow of the photographer. (White Sands National Monument)

Fig. 1.6C (RIGHT) Same as Figure 1.6B except that the shadow of the head, being cast on a more distant dune, shrinks.

otherwise!), our own shadow takes on a triangular outline due to perspective.

Strolling past the Kitt Peak solar telescope one morning a light fog or mist was invading at ground level. The sun shone brightly behind the structure casting sharply detailed shadows (Figure 1.6D). As this mist was closer than the telescope, the shadows appeared enlarged (Figure 1.6E). Then we spied an even more gigantic spectre in the overall mist.

Fig. 1.6D Double shadow cast on a thin fog layer, actually two layers, between us and the telescope. These layers being closer than the object, the shadows appear enlarged. This might be termed an 'anti-Brocken Spectre'.

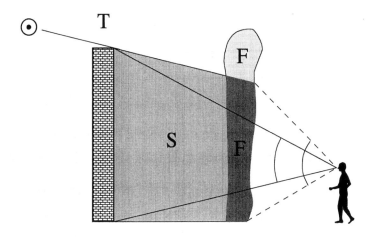

Fig. 1.6E The telescope's shadow S is cast on a thin fog F between us and the telescope T. The shadow on the fog is closer than the telescope and therefore subtends a larger angle, even though the umbra is tapering slightly and is therefore physically smaller than the telescope.

1.7 Spectre of the Brocken

Suppose we are on a hillside with fog billowing up to almost envelop us. At our back the sun shines brightly in a clear sky and our shadow penetrates this mist below for some tens of meters. Our shadows are transformed into apparently pyramidal silhouettes. Tapering to their apices at the antisolar point, our heads' shadows are small and indistinct. In fog, our shadows show rays converging to the antisolar point around our heads (Figure 1.7A). If we swing our arms, the shadows ape these motions. This is the Spectre of the Brocken, so named for the commonly reported phenomena seen in the Hartz mountains of Germany[7]. The spectre may be further enhanced by an accompanying glory or fogbow. Any companion's shadow is largely invisible to us because our oblique line-of-sight through his shadow is too small to produce any noticeable contrast.

A spectre is nothing more than our own three-dimensional shadow seen in perspective (Figure 1.7B). The length of the umbra may be no more than 30 meters but it seems larger. This probably accounts for the thrilling impact it has on the observer, who feels magnified and personally involved. Spectre-like shadows also can be encountered at night if we stand in front of an automotive headlight which is directed toward a fog bank. The resulting umbra grows without limit since the source of light is nearby and smaller than ourselves.

Fig. 1.7A Spectre of the Brocken. (Topanga, California)

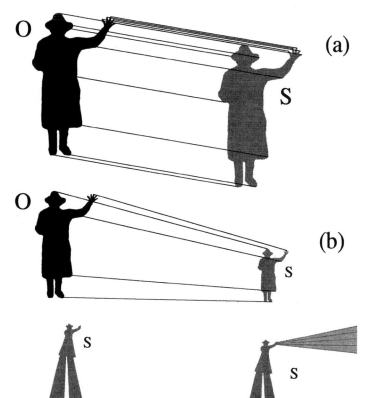

(a)

(b)

(c)

(d)

Fig. 1.7B A shadow appears different depending on how it is viewed and on what it is projected: (a) The umbral shadow (S) of the observer (O) is nearly the same size as the observer, here seen from the side. The source of light is low and the shadow is cast on a nearby vertical wall. (b) Another observer sees things differently when standing to one side and slightly toward the source of light. (c) The observer sees his own shadow stretched out in a roughly triangular shape when the shadow is cast upon the ground. (d) When the shadow is cast on fog, the observer sees it in three dimensions.

1.8 Mountain shadows

Should you find yourself atop a mountain as sunset approaches, put the sun at your back and look for the mountain's shadow. Regardless of the mountain's true profile, its shadow always appears triangular with the shadow's apex lying at the antisolar point (Figure 1.8A). Besides the ground shadow, the shadow cast on the atmosphere may be visible as well, especially if there is a significant amount of dust in the sky[8]. Light from the dust outside the shadow scatters sunlight. Inside the shadow, this component of scattered light is absent and the shadow appears dark.

What makes the mountain shadow triangular regardless of the mountain's true profile? The answer is perspective. Seen from above, the shadow is an accurate, though stretched version of the mountain's true shape (Figures 1.8B, 1.8F). But seen from the summit, the shadow appears to converge in the distance. Structures on the mountain's profile are reduced to minor and generally undetectable perturbations to the smooth edge of the shadow (Figure 1.8C). This also explains why railroad tracks, though parallel, always appear in perspective to be shaped like a triangle as they vanish in the distance.

The sharpness of the shadow is remarkable. Imagine yourself at the summit. When sighting along the edge of the shadow you are

Fig. 1.8A Mountain shadow of Kitt Peak viewed from the summit appears triangular even though the mountain itself is flat-topped. Details of the mountain summit profile are collapsed into the apex of the shadow and so are not resolved. (Kitt Peak, Arizona)

looking in the same direction as the sun's rays which define the shadow. Thus the apparent angular width of the penumbra is as small as possible, namely 1/2°. Since you are casting your own minute shadow along with the mountain, your line-of-sight coincides precisely with the direction of the sun's rays. From the summit the shadow appears greatly foreshortened and the edges of the shadow appear sharp.

If the observer is not on the summit but some way down the shady side of the mountain, he will see a spike stretching upwards from the shadow's apex (Figure 1.8D), usually tilting to one side or the other[9]. In very clear air the spike is absent. A spike appears because the observer is now entirely within the three-dimensional shadow. Since he is surrounded by it, his line-of-sight in any direction will pass through the shadow and into the sunlit regions beyond. The contrast of the shadow against the brighter sky will be proportional to the line-of-sight distance through the shadow (Figure 1.8E). A spike emerges from the observer's vantage point along the locus of maximum line-of-sight through the shadow. For an observer to the left of the summit, the shadowed line-of-sight is largest to his right and the spike is located there.

Fig. 1.8B View of Coyote mountain shadows. Note the severe elongation which tends to magnify summit structure. (Kitt Peak, Arizona)

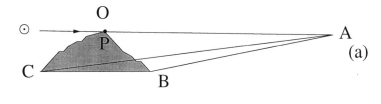

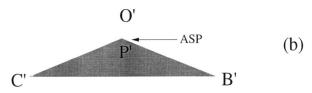

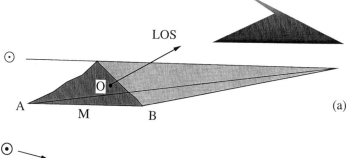

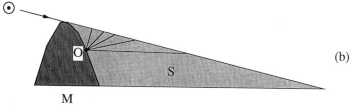

Fig. 1.8C Optics of mountain shadows when the observer (O) is on the summit. (a) The umbral volume ABCO is roughly the same size and cross section as the mountain in profile BCO, except that it tapers gradually to a point (A) owing to the sun's $\frac{1}{2}$ degree angular width. (b) From the summit the mountain shadow B'C'O' appears triangular as a result of perspective. O', the apex of the mountain shadow, is at the observer's antisolar point ASP.

Fig. 1.8E Optics of a mountain shadow spike as seen from off the summit. (a) The observer O in the shadow of the mountain sees a spike extending from the top of the mountain shadow. This spike points left if the observer is closer to B and to the right if he is closer to A. The spike is a long faint extension of the shadow that reaches (in principle) across the sky. (b) It is there because the observer's line-of-sight (LOS) through the shadow stretches from near the horizon to almost overhead.

Fig. 1.8D Mountain shadow spike. (Mt Popocatepetl, Mexico)

Fig. 1.8F Mountain shadows from over the world: Pic du Midi, France, J. Rösch; Pic du Midi, France, J. Rösch; Baboquivari, Arizona, B. Schaefer; Kitt Peak, Arizona, W.L.; Adams' Peak, Sri Lanka, A.C. Clarke; Mt. Blanc, France, P. Turon; Mt. Whitney, California, B. Cardon; Mauna Kea, Hawaii, B. Schaefer; Pico del Teide, Canary Islands, L. Cox; Humphrey's Peak, Arizona, W.L.; Mt. Fuji, Japan, anon.; Everest/ Qomolangma, Nepal/China, P. Athans.

1.9 Rays

They go by many names: crepuscular rays, sunbeams, sun drawing water, Rays of Buddha, or Ropes of Maui. All appear as light and dark shafts of light extending radially outward from the sun (Figures 1.1A, 1.1B). They can even be seen under water (Figure 3.7A). They are most prominent around sunset and sunrise (Figure 1.9E). These rays are called crepuscular (twilight) rays, although they can be seen when the sun is high as well[10]. Rays always point to the sun, even when it is below the horizon. They may be either light (against a darker background) or dark (against a lighter background). Sometimes both are present in a display.

Rays form when clouds cast their shadows on the air (Figures 1.9A, 1.9B). Bright rays happen when light shines between clouds

(Figure 1.9F). Dark rays are just shadows of clouds cast upon the air. A ray's contrast depends on many factors including the scattering angle to the sun, the line-of-sight distance through the ray and the number of scattering particles.

Light from the sky comes from a volume of air, each little bit of which scatters sunlight in our direction. The scattering may be from air molecules and dust particles. When a part of this volume is shaded, it scatters less light than its surroundings and thus may be visible by virtue of the contrast between it and illuminated space.

Rays achieve their highest contrast when viewed towards or away from the sun rather than at 90° to it. There are two reasons for this. First, forward scattering and backscattering by dust and

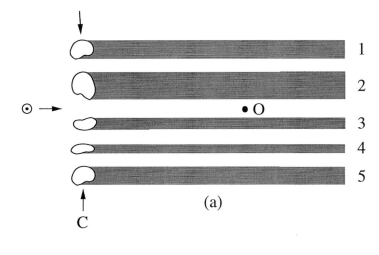

(a)

C

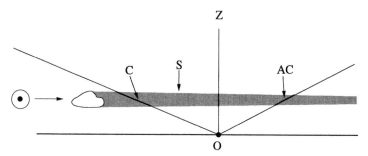

Fig. 1.9B A crepuscular ray viewed from the side. Note that the line-of-sight distance through the shadow S is smallest overhead and largest towards and away from the solar and antisolar points. (C crepuscular; AC anticrepuscular; Z zenith.)

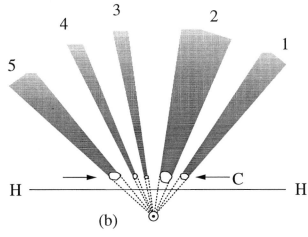

(b)

Fig. 1.9A Optics of crepuscular rays. (a) Looking down on the observer O from the zenith. Clouds C cast long straight parallel shadows 1,2,3,4,5. (b) When viewed in perspective, these shadows seem to diverge away from the sun (☉). The sun may be either above or below the horizon H but the geometry is the same.

air molecules are strongest in these directions. Second, the line-of-sight through the ray is largest when the observer is looking along a ray towards or away from the sun.

The convergence of rays is a matter of perspective. Although they are parallel (because the sun is effectively at infinity), the location of the observer between rays makes them appear to converge in much the same way that railroad tracks do. Rays from the

low sun may stretch for hundreds of kilometers through the sky (Figure 1.9C). Since the rays are long and parallel, we might expect to see them converging to the antisolar point on the opposite side of the sky. And we do. These so-called anticrepuscular rays are often overlooked (Figure 1.9D). Those portions of the rays lying overhead and connecting the rays on opposite sides of the sky are normally invisible because the line-of-sight through them is relatively short. Anticrepuscular rays glow with a lovely color contrast between the dark blue shadows and the pinkish sunlit sky.

In mid-latitudes (±20–50°), rays appearing in the west at sundown predict coming clouds. Most storms move in from the west and cumulus clouds over the horizon are revealed by rays in the clear western sky.

Fig. 1.9C Shadows cast by towering storm cells can be hundreds of kilometers long. Viewed from space we see these shadows undistorted by perspective. (Photo courtesy NASA)

Fig. 1.9E Late afternoon shadow rays from the Matterhorn, Switzerland.

Fig. 1.9D Anticrepuscular rays are a continuation of crepuscular rays reaching toward the antisolar point. (Oaxaca, Mexico) (Photo courtesy John Lutnes)

Fig. 1.9F Break in the clouds creates this crepuscular column of light. (Photo by C. Hunter)

1.10 Other shadow phenomena

Shadows play a prominent role in many of the optical phenomena we discuss elsewhere in this book; the earth's shadow, twilight, eclipses, eclipse shadow bands, shadows in water, aureole effect, rainbow wheels, whiteouts and the cloud contrast bow.

References

1 Goethe, J.W. von, 1970, (reprint) *Theory of Colors*, MIT Press, Cambridge.
2 Hapke, B. and Van Horn, 1963, 'Photometric studies of complex surfaces, with applications to the moon', *Journal of Geophysical Research*, **68**, 4545.
3 Irvine, W.M., 1966, 'The shadowing effect in diffuse reflection', *Journal of Geophysical Research*, **71**, 2931.
4 Trowbridge, T.S., 1978, 'Retroreflection from rough surfaces', *Journal of the Optical Society of America*, **68**, 1225.
5 Trowbridge, T.S., 1984, 'Rough-surface retroreflection by focusing and shadowing below a randomly undulating interface', *Journal of the Optical Society of America*, part A, **1**, 1019.
6 Hapke, B., 1986, 'Bidirectional reflectance spectroscopy 4. The extinction coefficient and the opposition effect', *Icarus*, **67**, 264–80.
7 Flammarion, C., 1874, *The Atmosphere*, Harper & Bros., NY.
8 Livingston, W. and Lynch, D., 1979, 'Mountain shadow phenomena', *Applied Optics*, **18**, 265.
9 Lynch, D.K., 1980, 'Mountain shadow phenomena. 2: The spike seen by an off-summit observer', *Applied Optics*, **19**, 1585.
10 Lynch, D.K., 1987, 'Optics of sunbeams', *Journal of the Optical Society of America*, part A, **4**, 609.

2 Clear air

From the mast-head the mirage is continually giving us false alarms. Everything wears an aspect of unreality. Icebergs hang upside down in the sky; the land appears as layers of silvery or golden cloud. Cloud banks look like land, icebergs masquerade as islands or nunataks, and the distant barrier to the south is thrown into view, although it really is outside our range of vision. Worst of all is the deceptive appearance of open water, caused by refraction of distant water, or by the sun shining at an angle on a field of smooth snow or the face of ice-cliffs below the horizon.

Sir Ernest Shackleton in *South!*, from the Captain's log of the exploration ship 'Endurance'[1].

Planet earth is blessed with a remarkable envelope of air; no other planet has such a changeable sky. Our atmosphere is thick enough to carry water aloft and form clouds. Yet it is thin enough to be transparent in the visible part of the spectrum and to allow the sun and stars to shine clearly down upon us. If the amount of air were doubled or halved, the sky would most likely cease to be interesting, becoming either opaque from a perpetual overcast or clear and featureless. If the earth was but a few percent closer to or farther from the sun, the abundant surface water would either evaporate entirely or freeze out as ice. Only on earth can our atmosphere, whose load of water vapor hangs in a delicate balance between transparent and opaque, sport the amazing variety of optical phenomena seen as the sun goes about its daily rounds[2].

2.1 Sunlight and atmospheric absorption

On a clear sunny day it seems that the sun is shining with all its might. But it's not. Much of the sun's light never reaches to the ground (Figure 2.1). Solar radiation striking the earth above the atmosphere is attenuated in passing through the air by two processes: absorption and scattering. Absorption removes light from the beam of light and converts it to heat. Absorbed light is never seen again in the visible part of the spectrum, though it may appear as thermal radiation far into the infrared. Absorption does not occur uniformly across the spectrum, but only at certain

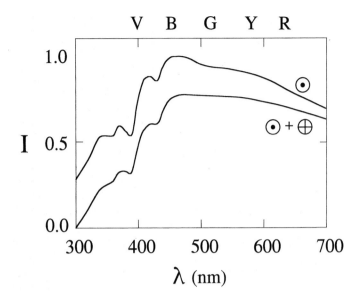

Fig. 2.1 Sunlight outside the atmosphere and at the ground. The upper curve (⊙) is solar radiation and the lower curve (⊙+⊕) is sunlight as seen by a sea level observer with the sun at 20° altitude. The spectral resolution of both curves is too low to show individual absorption lines.

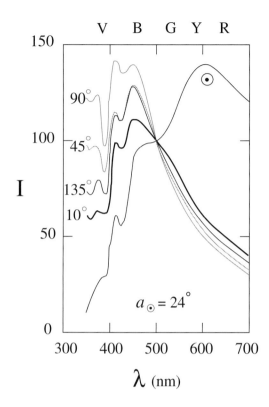

Fig. 2.2A Comparison of the spectra of sunlight and skylight showing that the color of skylight depends on the distance in the sky from the sun[9]. The sun is at an altitude of 24°. The spectra were taken on the solar vertical circle at angular distances from the sun of 10°, 45°, 90° and 135° respectively. All spectra have been scaled to have the same value at 500 nm (Henderson[3] after Hess[9]).

discrete wavelength regions determined by the absorbing molecule's internal properties. Scattering, while not absorbing energy, redirects it out of the beam and away from its original direction. It takes place at all visible wavelengths. The agents responsible for both processes are air molecules and dust particles[3]. The light of the sun we see from the surface of the earth is called sunlight. The combined mixture of sunlight, skylight (light scattered from the air and clouds) and light reflected from the ground we call daylight.

2.2 The clear blue sky

The sky is blue, right? Yet how many of us recognize these aspects?

(1) Color – The sky actually contains all colors. Near the horizon it is nearly white, but it may be tinted one of several colors due to reflection from the ground.

(2) Brightness – It is faintest at the zenith and rapidly brightens near the horizon (for a mid-day sun). The sky is also darker at higher elevations.

(3) Polarization – Light from the sky is polarized, varying between around 85% at 90° from the sun to zero at other places. There are four small unpolarized regions on the solar vertical circle called neutral points.

To understand the color of the sky, we must first consider sunlight itself. Sunlight consists of light of every wavelength and polarization plane. Its spectrum is shown in Figure 2.2A. This unpolarized light is very close to perfect white. It reaches the earth and begins filtering through the atmosphere.

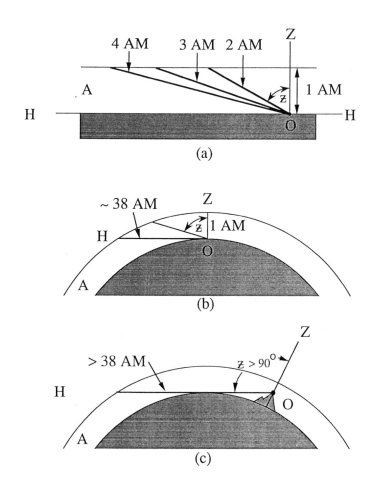

(a)

(b)

(c)

Fig. 2.2B Air mass geometry. (a) The air mass, AM, is defined as the amount of air A through which the observer O is looking. The amount of air is measured with respect to the zenith distance z and varies roughly as secant z. AM is expressed in units of 1 air mass = amount of air in the vertical direction (zenith distance $z = 0°$) at sea-level. (b) For values of $z > 60°$, the AM = secant z is not valid and the spherical properties of the atmosphere must be included. On the horizon H the air mass is about 38. (c) When the observer is above sea-level the air mass exceeds 38 because the line-of-sight to the tangent point of the earth is below the local horizon and therefore passes through more atmosphere.

According to Lord Rayleigh's theory of molecular scattering[4–8] the probability that a single photon of sunlight will be scattered from its original direction by an air molecule is inversely proportional to the fourth power of the wavelength. The shorter the wavelength (or 'bluer') the light is, the greater its chances are of being scattered. For example blue light, with a wavelength of 450 nanometers, is 3.2 times more likely to be scattered than red light with a wavelength of 600 nanometers ($(600/450)^4 = 3.2$). This means that when we look in any part of the sky except directly toward the sun, we are more likely to see a blue photon of scattered sunlight than a red one. This causes the sky to appear blue. Figure 2.2A compares the spectral distribution of sunlight and skylight[9]. Note that even the bluest sky still contains all colors of the spectrum.

The brightness of the sky is determined by the number of molecules in the line-of-sight: more air molecules mean a brighter sky. From a mountain or a high-flying jet the sky is darker than it is seen from the ground. This is because there are fewer air molecules in the line-of-sight. In space, or on the moon, there are no molecules and therefore no scattered light to brighten the sky. There the sky is black.

Light traverses the minimum amount of atmosphere when its path is perpendicular to the surface, i.e. when it comes from the zenith Z (Figure 2.2B). This quantity of air is defined as 1 air mass. At larger angles from the zenith (denoted by zenith 'distance' z), the air mass increases roughly as the secant of the zenith distance. As a result, the sky brightness increases to a maximum just above the horizon.

If the earth and its atmosphere were flat, the air mass would be exactly proportional to the secant z. But owing to the earth's curvature, the air mass, and therefore the absorption and scattering fall slightly below secant z. On the horizon at sea level sunlight passes through about 38 air masses. Figure 2.2C shows how the air mass depends on zenith distance.

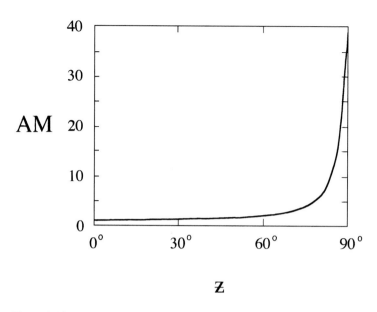

Fig. 2.2C Air mass, AM, as a function of zenith distance z. Because the air mass varies roughly as secant z, it is near unity at small zenith distances and slowly changes until zenith distances of greater than about 45°.

2.3 The sky near the horizon

During the day, the horizon sky is much brighter than it is near the zenith. It is also nearly white. Why?

Towards the horizon the line-of-sight passes through many more air masses than when looking at the zenith (Figures 2.2B, 2.3A). Since each particle is a source of scattered light we might think that by increasing the number of scatterers, an arbitrarily bright horizon sky could be made. But this is not the case.

The brightness of the horizon sky is found to be about as bright as the sky can be. The addition of further air would not brighten it further (Figure 2.3B). This is because when there are enough air molecules in the line-of-sight, the sky becomes almost opaque due to multiple scattering, i.e. scattering of light more than once and possibly many times. The process is similar to how the ground looks as snow begins to fall. At first the ground is dark but slowly brightens as each snowflake covers a bit of it. When the ground is covered, the addition of more snow does not cause further brightening because the layer of snow is now opaque. Snow, like the air,

does not absorb much light and therefore its opacity is due largely to scattering.

Opacity is measured in terms of optical depth τ (greek letter tau) and is proportional to the number of particles. Something that is nearly transparent like a piece of window glass or the overhead sky is said to be optically thin, with τ being much less than one. Opaque objects have very large optical depths (τ >> 1). The cross-over point between optically thick and thin occurs when τ = 1, at which point the medium transmits only about 37% of the light incident upon it.

As mentioned before an object can remove light from an incoming beam in one of two ways: absorption and scattering. Absorption converts incoming light into heat or other internal molecular energies but scattering simply sends it into a different direction. When scattering is the dominant source of opacity, the light can be seen coming from some direction other than the

Fig. 2.3A All-sky photograph showing that the sky near the horizon is brighter than the sky overhead. Also note the solar aureole (Section 2.7) around the hand of the photographer as he blocks the direct sunlight.

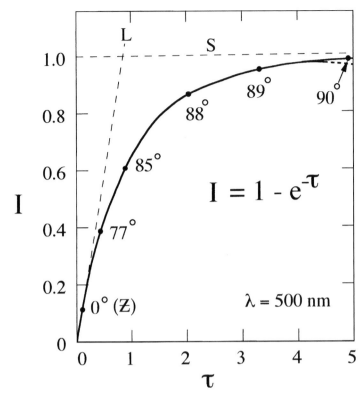

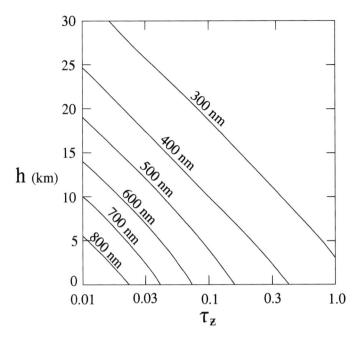

Fig. 2.3C Optical depth τ measured vertically from various elevations *h* and wavelengths. The optical depth is greater for shorter wavelengths because Rayleigh scattering is more efficient at these wavelengths.

Fig. 2.3B Relative brightness *I* of the sky as a function of optical depth τ (thick curve) at a wavelength of 500 nanometers. For small optical depths the brightness is proportional to τ and falls on the linear (L) part of the curve. At greater optical depths the brightness increases more slowly and ultimately reaches a maximum value. At this saturation level (S) further increases in the optical depths do not lead to increased brightnesses. The zenith distances of various optical depths are shown on the thick curve.

source. In the case of the horizon atmosphere, the light we see is multiply scattered sunlight.

Why does multiple scattering cause the atmosphere to lose its blue color near the horizon? We have already seen that in an optically thin atmosphere where only single scattering takes place, the sky is blue. But in an optically thick scattering medium, all of the colors are scattered many times before reaching the observer's eye. Molecular scattering is still at work and the longer wavelength photons still have the least probability of being scattered. But eventually all photons, regardless of their wavelength, strike air molecules. They scatter and rescatter so many times that they all get thoroughly mixed together. Since there is little or no absorption in molecular scattering, the sky near the horizon is the same color as the sun, i.e. white.

It is easy to calculate the altitude at which the sky starts to become white. Suppose that the zenith optical depth is about 0.1, which is about right for green light (Figure 2.3C). Optical depth is proportional to the number of scatters in the line-of-sight and therefore the optical depth must be 1 when the air mass is 10 (0.1×10 = 1). This occurs near 5° altitude. Above 5° the sky should be distinctly blue because the optical depth is small. Below 5° it should be bright and nearly colorless and be completely white at the horizon where the optical depth is about 3.8 (0.1×38 air masses).

The whiteness of the horizon sky may be tinted by light reflected from the landscape. Over water the low sky is dark, over vegetation it becomes slightly greenish, and in the desert it is brownish-yellow.

When looking at a sunset from a mountain, our line-of-sight passes through more atmosphere than it would if we were at sea level, not less (Figure. 2.2B). This may seem surprising since there is less atmosphere above us than there is at sea level. But from a mountain, the setting sun is not on the horizon ($z = 90°$) but rather it is below the horizon ($z > 90°$). The additional atmosphere lies between us and the point on the earth where our line-of-sight is tangent to the surface. Beyond that, there is another 38 air masses to the sun. From space, the setting sun is seen through $2 \times 38 = 76$ air masses, a fact demonstrated by the increased distortion in the rising moon as seen from space[10].

Having just discussed why the sky brightens toward the horizon, we note that recent measurements show that the sky actually darkens slightly just above the horizon[11]. This is a result of the sky's significant optical depth. In the visible at 500 nanometers (Figures 2.2C, 2.3C) when we look toward the horizon at sea level we are looking through at least 38 air masses which corresponds to an optical depth of 6, and usually more owing to the presence of aerosols. As a result, light scattered toward us from the horizon is dimmed slightly and the horizon dims. It is a small effect (dotted line in Figure 2.3B) as evidenced by the fact it was only discovered in 1994!

That there is any reversal in the brightness vs. altitude is a quirk of our atmosphere. Were it thinner, the sky brightness would increase monotonically toward the horizon. Were it much thicker, the sky would be uniformly bright at all scattering angles, as it nearly is on an overcast day.

2.4 Sky polarization

The blue sky is polarized. The plane of maximum polarization is roughly 90° from the sun (Figure 2.4A). Away from this plane the polarization decreases. This fact is well known to photographers who often darken the sky by adjusting their polaroid filters to minimize sky light.

The sky is polarized because air molecules scatter light of different polarizations in different ways. Light that is scattered a single time by 90° (i.e. in the 90° scattering plane) is almost totally polarized. At less than 90°, it is partially polarized and reaches a minimum value near zero at 0° and 180°. Light scattered at 90° is not 100% polarized because air molecules are not perfect scatterers. At the horizon where multiple scattering takes place, all polar-

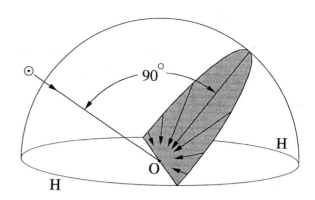

Fig. 2.4A The sky is most strongly polarized 90° from the sun. Maximum polarization in the plane is about 80%.

izations are present in roughly equal amounts and the sky is nearly unpolarized.

Air molecules scatter about the same amount of light into the forward hemisphere as they do into the backward hemisphere. The theoretical scattering pattern for a perfect air molecule, one that is small compared to the wavelength of light, is shown in Figure 2.4B. This scattered light reaches a brightness minimum at the 90° scattering angle. This minimum in the plane perpendicular to the incident light is due to the absence of one polarization component. This minimum is called the band of darkness[12]. It is a broad, shallow brightness minimum coinciding with the plane of maximum polarization.

Pure molecular scattering would suggest that the sky should be 100% polarized 90° from the sun. But observations show that it reaches only 75 to 85% at most. This lack of full polarization occurs for two reasons. First, air molecules are not perfect scatterers and as a result their scattering patterns depart slightly from those shown in Figure 2.4B. Second, multiple scattering causes a certain amount of light to strike the molecule from directions other than from the sun. The sources of this light are the sky itself and the ground. When all this multiply-scattered light is added to the light scattered directly from the sun, it reduces the amount of polarization.

Observers have found several unpolarized regions in the clear sky that should be polarized if only single molecular scattering

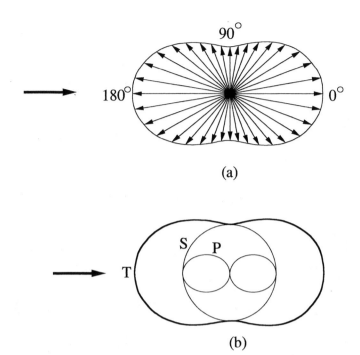

(a)

(b)

Fig. 2.4B (a) Scattering by an air molecule is symmetric: As much light is scattered into the forward plane (towards 0° scattering angle) as in the backward direction (towards 180° scattering angle). The least amount of light is scattered into the 90° direction and this defines the plane of maximum polarization and minimum brightness in the sky. (b) This occurs because one component of polarization, the perpendicular component P is zero in the 90° scattering direction. Since the total (T) scattering function is the sum of the parallel (S) and perpendicular (P) components, where one polarization is absent, the total component is most polarized and least bright. In the forward and backward directions T is at a maximum and the polarization is zero.

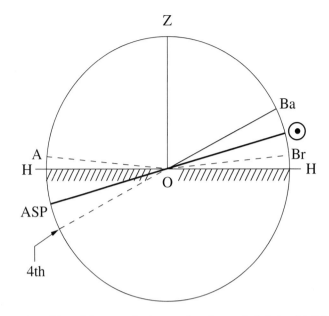

Fig. 2.4C Position of the neutral points on the solar vertical circle with Z the zenith. At these points the sky shows no polarization. The neutral points are named for their discoverers: the Brewster point, Br, 10–20° below the sun, the Babinet point, Ba 15–20° above the sun, and the Arago point, A, 20° above the antisolar point ASP. A fourth neutral point has been postulated that should be directly opposite the Babinet point. Since this point is normally below the horizon, it cannot be seen except from high altitude.

was at work[13,14]. These so-called neutral points all lie on the solar vertical circle i.e. the great circle passing through the sun, zenith and nadir (Figure 2.4C). Named for their discoverers, the neutral points are the Brewster point 10–20° below the sun, the Babinet point 15–20° above the sun, and the Arago point 20° above the antisolar point. A fourth neutral point should occur directly opposite the Babinet point. Since it is normally below the horizon, it cannot be seen except from high altitude and to our knowledge it has never been reported. Other neutral points are sometimes observed, especially when the sky is heavily laden with aerosols[15]. The position of the neutral points also changes when viewed over water[16], presumably due to the addition of light scattered from the water.

Neutral polarization points occur because the sky is illuminated by two sources of light: the sun and the bright horizon sky. Even though the intensity per square degree of the horizon sky is small compared to the sun, it subtends a very large angle (roughly 1500 square degrees compared to the sun's 0.2 square degrees) and is a significant source of light. Due to scattered light from the ground, horizon skylight is slightly polarized. When this light is added to the horizontal component of polarized skylight from the direct molecular scattering of sunlight, certain areas are rendered unpolarized.

Fig. 2.5A Airlight is a thin veil of light interposed between the observer and
(say) a mountain.

2.5 Airlight

Distant mountains look bluish. And the further away the mountains are, the bluer and brighter they appear (Figure 2.5A). This blue veil is called airlight. Airlight is most evident when the sky is clear and is seen against distant objects.

Airlight is simply sunlight that is scattered by air molecules between us and the mountain. In fact skylight is just airlight, but since it is not seen against anything except dark space, it is unremarkable. The amount of airlight depends on the distance to the mountains. In the limit of a very distant mountain, multiple scattering renders the atmosphere opaque. Light from the mountain is scattered out of the line-of-sight and replaced by airlight. When this happens the mountain vanishes. Thus in even the purest of air, there is a limit to how far we can see.

Haze can masquerade as airlight. Unlike true airlight which is blue, haze appears gray, white, or brown. With their strong forward scattering component, dust and other aerosols make airlight that is very bright in the direction of the sun (see aureole Section 2.7).

Successive ridges of mountains reveal airlight at its best. As the eye scans upwards, the distance through the air changes abruptly from the ridge of one mountain to the face of the next (Figure 2.5B). At this point there is a step change in the amount of airlight and thus a contrast change[17].

When looking towards the sun, airlight can be so bright that distant mountains cannot be seen. After sunset, however, mountains that were previously invisible pop into view (Figure 2.5C). When the sun is up it creates an enormous amount of airlight. This lowers the contrast of the mountain to the point of invisibility. But when the sun goes down the airlight decreases sharply and the mountain becomes visible as its contrast with the surrounding sky increases.

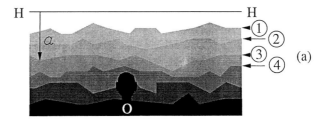

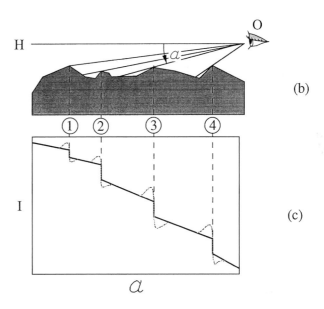

Fig. 2.5B Airlight geometry and Mach effect in the image of a series of mountain ridges. (a) An observer O looking at a series of successively more distant ridges sees that those nearest the horizon [smaller a (b)] are brighter than the nearer ones (c). This is because they are farther away and therefore the intervening airlight is brighter as a result of their greater distance (b). In such a situation, the Mach effect creates a common optical illusion. The observer has the impression that instead of simply having abrupt changes in brightness as the eye scans from ridge to ridge as the solid line in (c) shows, there are bright and dark bands (Mach bands) adjacent to the transitions.

Fig. 2.5C Upper photo is taken mid-day looking across the Gulf of California toward Baja from Puerto Peñasco, when airlight is strong. Lower photo is looking in the same direction after sunset, and the distant mountains of Baja are now clearly visible in profile owing to the reduction of airlight. Note that the shorelines differ because of tides.

2.6 Color and brightness of the low sun

As the sun approaches the horizon, its color changes from dazzling white to bright yellow, orange and even to deep red. No two setting (or rising) suns are alike, yet they all have one thing in common: they are dimmer and redder than when they are overhead.

When the sun is high, the amount of light removed from a beam of sunlight is small because the optical depth of the sky is small. The sun appears very nearly the same color as it does outside the atmosphere: white. When low, however, its light passes through many times the number of air molecules than at the zenith and a significant fraction of its light is scattered out of the beam. From molecular scattering theory we know that this is especially true near the blue end of the spectrum. With a large fraction of its shorter wavelengths missing, the sun appears yellowish (Figure 2.6A). In addition to scattering light, ozone absorbs heavily in the blue and green part of the spectrum, thereby further reddening the low sun.

But there is another factor that is needed to explain red suns: particles. These are primarily dust and smoke, though over the ocean minute water droplets are also present. Particles smaller than about 100 nanometers in diameter are particularly efficient scatterers and their influence is far greater than that of air molecules. Their presence in the atmosphere further reddens the sun. Because the number of particles is variable from day to day, successive sunsets are equally variable.

Light from a distant mountain, like light from the low sun, is attenuated and reddened by the atmosphere. Because reddening is accompanied by attenuation, most mountains simply fade away and they can only be seen as low contrast blue ridges as airlight replaces the mountain's light. But when the mountain is snow capped and therefore bright, it remains bright enough to be clearly visible even after considerable reddening and attenuation and can often appear slightly yellowish (Figure 2.6B). Distant clouds frequently look yellowish or orangish. Sometimes, when conditions are just right, a veil of blue airlight mixes with the reddened light from the mountain to reveal the 'purple mountain's majesty'. Thus scattering and absorption can make distant mountains look almost any color: bluish (if they are dark) or yellowish (if they are covered with snow).

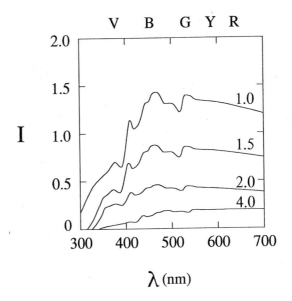

Fig. 2.6A Change of color of sun with altitude (Ref. 3, Figure 60). These spectra show the sun as viewed through 1.0, 1.5, 2.0, and 4.0 air masses. Note that with increasing air mass the sun becomes dimmer and redder.

Fig. 2.6B Moods of Mt. Everest (Qomolangma), as seen from Rongphu Monastery, Tibet. (left) Normal sunlight with cap cloud hovering over summit, (center) illumination by a low sun, (right) alpenglow conditions (Section 2.13) well after sunset. Note the absence of shadows.

2.7 The aureole

That intensely bright glare surrounding the sun is called an aureole. It is usually a few degrees across and fades rapidly away from the sun (Figure 2.7A). It is the same color as sunlight (white) and can vary enormously in both brightness and size. The aureole is extremely bright and nearly impossible to look at directly. Photographing it is even more difficult: not only do most films lack sufficient dynamic range ('latitude'), ghost images of the sun and scattered light in the camera mask the aureole or at least prevent an accurate picture from being recorded. On rare occasions, usually in the mountains when the air is exceptionally clear, there is no aureole (Figure 2.7B).

Fig. 2.7A (LEFT) Weak aureole around the sun. The sun is hidden by a street lamp. To the eye, the sky appeared perfectly clear.

Fig. 2.7B (RIGHT) The next day the sky was exceptionally clear and there was no aureole, indicating that virtually no aerosols were present.

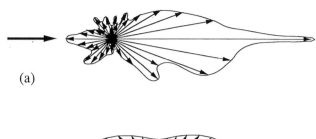

(a)

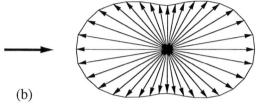

(b)

Fig. 2.7C Scattering patterns for different particles. (a) Large irregular particles, like those comprising dust and smoke, are irregular in the sense that they are not symmetric. They do, however, have a strong forward scattering peak and a smaller though still pronounced backscattering peak. (b) Air molecules have a scattering function that is symmetric fore and aft: they scatter the same amount of light in both the forward and backward directions but lack both the forward and backscattering peak. (c) Large water drops have a strong forward and backscattering peak and also show strong enhancements at the primary and secondary rainbow angles.

The aureole originates from forward scattering by atmospheric particles, typically dust, smoke, tiny water droplets, smog particles, and pollen. Sometimes insects are numerous enough to create an aureole, though the presence of a few gnats at ground level might never suggest that they are responsible for the bright glow around the sun.

Each type of particle scatters light in a different way because they have different shapes. But they all have one thing in common: a strong forward scattering peak (Figure 2.7C). Though the explanation of this peak is somewhat complicated, it's origin is found in the fact that particles are made of millions of molecules, rather than single molecules. Air molecules play no part in the aureole because as Rayleigh scatterers they have no forward scattering peak.

2.8 Bishop's ring

The Bishop's ring is a faint bluish aureole-like disk of light encircling the sun. It is bounded by a pale narrow ruddy ring. The inner portions range from between 10° and 25° across. Owing to its location around the sun and its low contrast, the Bishop's ring is seldom noticed and even less frequently photographed. It was first reported after the eruption of Krakatoa in 1883[18,19].

The Bishop's ring is caused by scattering from stratospheric dust, almost always associated with volcanic eruptions. The particles have a fairly uniform size in the range of 1 micrometer[20]. Some of the particles may also be composed of sulfuric acid tetrahydrate[21] ($H_2SO_4 \cdot 4H_2O$). It is this restricted particle size that sets the Bishop's ring apart from the ordinary aureole and provides the color. Somewhat analogous to the normal cloud corona (Section 4.13), the ring shows low contrast diffraction colors with blue on the inside and red on the outside. Unlike a corona, there are never second order colors. Since fine volcanic dust spends many months in the stratosphere before gradually drifting earthward, the Bishop's rings are usually present for quite some time after a major eruption. Finally, the dust floats down to the troposphere where rain brings it to the ground.

TWILIGHT

2.9 Twilight

Twilight is that period of time when the sun is below the horizon yet continues to illuminate the sky. The sky, in turn, illuminates the landscape. Civil twilight begins at sunset and ends when the sun is 6° below the horizon. At this time outdoor reading becomes impossible. Nautical twilight ends when the sun is 12° below the horizon at which point only the gross outlines of objects can be discerned. Finally astronomical twilight ends at −18° and marks the onset of night. Dusk is twilight following sunset; dawn is the twilight preceding sunrise.

2.10 Guide to twilight

Many twilight phenomena occur within about 30° of the horizon because the air mass grows rapidly below 30°; that is where most of the air is. Owing to the symmetry between sunrise and sunset, the colors are the same in both events, though naturally occurring in the reverse temporal order. Figure 2.10A shows the celestial sphere along with the two most important sources of sky illumination – the low sun and sunlight scattered from the atmosphere.

The normal course of events at sunset goes something like this[12-14, 22-5]: With the solar disk on or just below the horizon, the western sky takes on a yellow or orange glow known as the twilight arch. Centered on the sun and the twilight arch is the aureole. To the east the bluish earth shadow rises. Over the next few minutes the deep blue earth's shadow, with the pinkish antitwilight arch as its upper bound, achieves its greatest contrast. While to the west, clouds or mountains lying beyond the horizon may cast their shadows to create alternating pink and blue radial stripes directed toward the subhorizon sun. These are crepuscular rays. (If extending into the eastern hemisphere of the sky, these features become anticrepuscular rays.) About 30 minutes after sunset (at a latitude of about 30°), the purple light may develop as a diffuse circular patch at 45° altitude over the sun (which is now well below the horizon). In the east, mountains or buildings begin to assume the warm orange tones of the alpenglow (Section 2.13). The purple light in the west fades in concert with a fading twilight arch. Finally it is dark (Table 2.10).

Table 2.10. *Sunset twilight sequence as seen from a latitude of 30° under a clear sky.*

Solar elevation (deg)	In the east	In the west
+5	horizon sky faint pink alpenglow yellowish	horizon changes from normal white to yellowish twilight arch
0	bluish earth shadow rising, capped by a pink antitwilight arch	sun setting, bright twilight arch
−2	earth shadow distinct, broadening antitwilight colors brightest	location of sun difficult to specify, twilight arch bright
−5	earth shadow becoming diffuse, antitwilight arch fading rapidly	purple light strong, twilight arch yellow/orange
−6	civil twilight ends, earth shadow gone, alpenglow purplish, antitwilight arch gone	purple light fading, twilight arch orange/red
−8	alpenglow gone	red twilight glow which spans large fraction of horizon fading
−12	nautical twilight ends	dim red twilight arch
−18	astronomical twilight ends, zodiacal light visible	

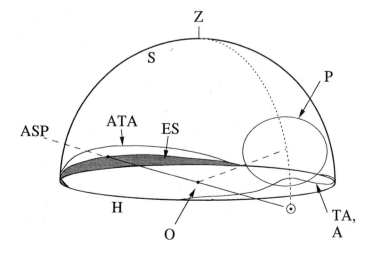

Fig. **2.10A** Celestial sphere and twilight phenomena. O – Observer, H – horizon, ⊙ – sun, Z – zenith, ASP – antisolar point, dashed line – solar vertical circle, ES – earth shadow, TA – twilight arch, A – low sun aurole, ATA – antitwilight arch, P – purple light.

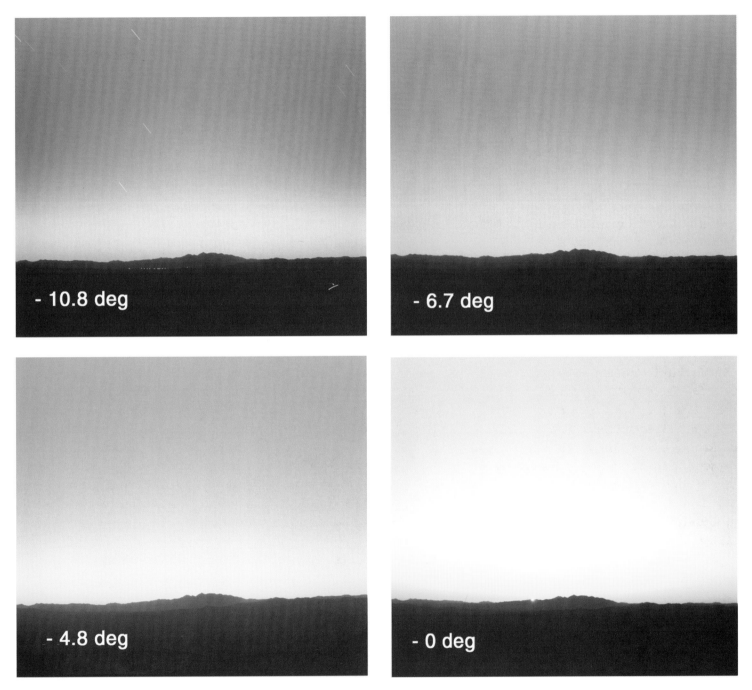

Fig. 2.10B Time sequence of the morning twilight colors. At a solar depression of −10.8° the stars still are visible but on the horizon there is a dim reddish twilight arch. By −6.7° the twilight arch has strengthened to an orange-red. At −4.8° the twilight arch is now orange-yellow and at its maximum and the purple light is present. With the sun on the horizon at −0° the twilight arch remains strong but the sky has brightened.

HOW TO PHOTOGRAPH THE TWILIGHT

It is difficult, if not impossible, to record photographically what our eye–brain perceives during a twilight event. At dawn the rested eye, with its tremendous capacity for adapting to different illumination levels, nicely transforms the twilight sequence. Cameras and films have limitations that one should be aware of.

If we wish a wide field we employ a short focal length camera. The distance between the film and the lens increases toward the edge of the field causing loss of sensitivity and the picture darkens markedly there. Film response is limited by what is called 'latitude'. The result is that if the exposure is correct for the central part, the edges will be irrecoverably dark, even black (see Figure 2.14A).

With these factors in mind we suggest a medium focal length camera and the use of print film. The latter has better latitude than the transparency type. Figure 2.10B is our best effort to record and represent what the eye sees. Considerable 'dodging' was necessary to achieve the flatness of intensity that is natural to the eye.

2.11 Twilight arch

That band of yellowish light that lingers low on the western horizon at sunset is called the twilight arch (Figure 2.11A). Stretching about 90° away from the sun on either side, its brightness diminishes horizontally away from the sun (the solar point). After sunset, the azimuth of the solar point may be difficult to identify. On other occasions the inner twilight arch is more akin to an aureole (Section 2.7), and is clearly located over the solar point by a localized glow that imparts an 'arch'-like structure.

As the sun retreats below the horizon the zenith sky darkens and the contrast of the twilight arch grows. It takes on a straw color and at the intersection between this yellowish band and the encroaching blue sky there may develop a layer having a greenish or turquoise tint. More often, however, there is no green color but just a merging of the yellow into blue. This period of maximum color contrast is both attractive and short-lived. Later, the vertical extent of the twilight arch decreases and the arch becomes red.

The twilight arch is sunlight scattered by the atmosphere (Figure 2.11B). Though the sun has set on us, it still shines on the air above us. Since the sun is low, it is yellowish and so is the twilight arch. As the sun gets closer to the horizon as seen by the sunlit air, it grows redder from Rayleigh scattering and this red light is scattered to us. The coppery red is reminiscent of the color of some lunar eclipses, and for good reason; their origins are the same. Why does the twilight arch hug the horizon? Because the air mass is high there and there are more air molecules to scatter the light to us than there are overhead.

Fig. 2.11A Twilight arch. Spanning the eastern horizon from Kitt Peak approximately 20 minutes prior to sunrise. Notice that the location on the horizon just above the solar point cannot be identified.

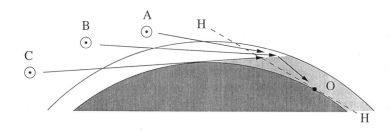

Fig. 2.11B Optics of the twilight arch. The twilight arch is due to sunlight that has been reddened by the atmosphere and scattered into our line-of-sight. This happens when the sun has set from the observer's O standpoint but still shines on the upper atmosphere and more westerly (at sunset) atmosphere. As the sun sets (A,B,C) the sunlight passes through more atmosphere and grows progressively redder. As a result the twilight arch grows fainter and redder as twilight deepens.

2.12 Earth shadow and the antitwilight arch

Just before sunset, a low flat, dark blue band rises up from the eastern horizon. This is the earth shadow and it stretches for nearly 180° (Figure 2.12A). It is bounded above by the pinkish antitwilight arch and below by the horizon. The earth's shadow is best seen when the sky is clear and our line-of-sight is long. From high elevation, the shadow appears sharper than it does from the ground. As the sun sets, the boundary between the earth shadow and the antitwilight arch rises in the sky and becomes progressively less distinct. With the disappearance of the antitwilight arch, the shadow blends smoothly with the deepening blue night sky. The earth shadow can be seen at twilight on most clear evenings. But for reasons not known to the authors, the blue apparition sometimes does not appear. Of course the shadow must be there but its visibility is low. Naturally, the earth shadow can be seen at sunrise, sinking into the west.

Figure 2.12B shows the geometry of the earth shadow. For a shadow to be visible, it must be cast upon something. In this case the earth shadow is cast upon the atmosphere itself. As the sun sinks below the horizon the shadow rises. The boundary of the shadow, the so-called twilight ray, is marked by the pink antitwilight arch. As the shadow rises, its boundary with the antitwilight arch broadens and grows less distinct. This happens because the observer's vantage point becomes more and more oblique through the boundary between shadow and antitwilight arch. The shadow boundary not only widens but also begins to accelerate upwards, seeming to rise faster than the sun sets[26]. Eventually the antitwilight arch fades to blue. Its strong color and brightness near sunset originate in backscattering by the relatively thick lower atmosphere. But as the twilight ray rises, so does the lower bound of the atmosphere, being illuminated by direct sunlight. Being

Fig. 2.12A Earth shadow at its maximum contrast.

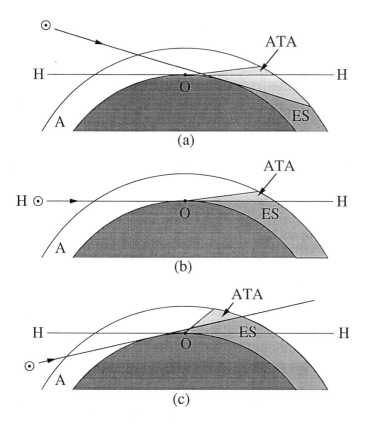

(a)

(b)

(c)

Fig. 2.12B Geometry of the earth shadow. The earth shadow ES, appearing as a low, dark blue band bounded above by the pinkish antitwilight arch, ATA, rises over the eastern horizon. (a) The shadow of the earth can be seen even before sunset because the atmosphere casts a shadow of itself. (b) At sunset the true earth shadow begins to rise and the antitwilight arch brightens and contrasts beautifully with the blue earth shadow. (c) After sunset the shadow of the earth rises and eventually blends with the antitwilight arch as twilight deepens.

Fig. 2.12C Earth shadow from a 'polar flight' showing the southern extremity of the shadow wedge. The object to the left is a waning gibbous moon, which has a declination of +13° for this date so the entire shadow is north of the equator. (Photo by Alan Clark)

higher and thinner, it scatters less red light and a point is reached where the strong bluish airlight, the same multiply scattered sky-light making the earth shadow blue, becomes the dominant source of light reaching our eyes (Figures 2.12C, 2.12D).

That the shadow can be seen before sunset would seem geometrically impossible. Yet its dark bluish rim can be seen several minutes before the sun disappears. What the observer is seeing is the shadow of the translucent atmosphere, behaving like a piece of smoky glass and casting its own indistinct shadow on itself.

Speaking of the color of the sky after sunset, why does the sky overhead remain blue during the entire twilight sequence when virtually every other part of the sky has changed color? The earth's stratospheric ozone layer (about 12 kilometers up) absorbs strongly at longer wavelengths, thus filtering out the reds, oranges and yellows leaving the zenith sky blue[27,28].

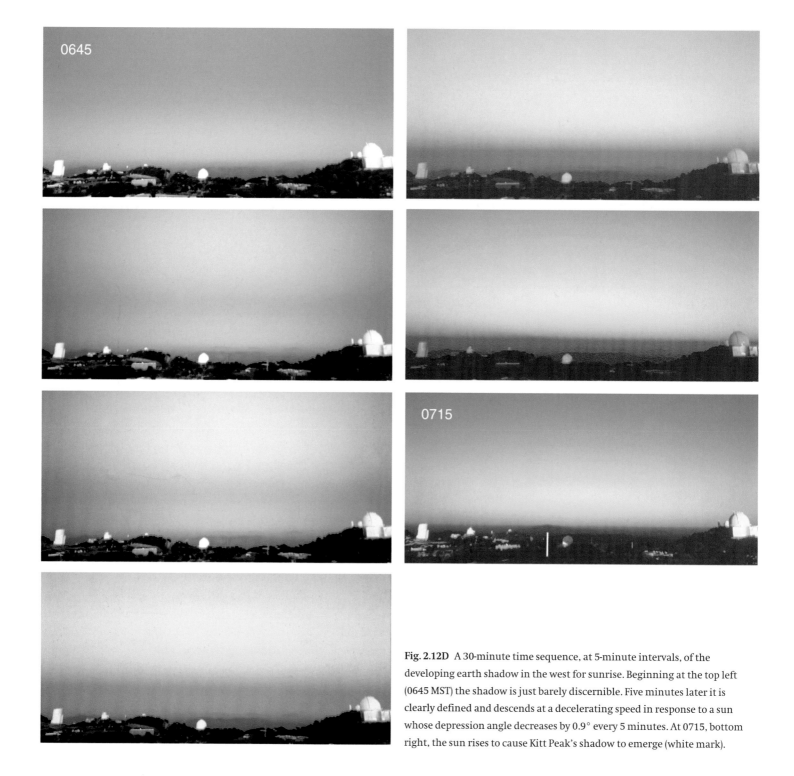

Fig. 2.12D A 30-minute time sequence, at 5-minute intervals, of the developing earth shadow in the west for sunrise. Beginning at the top left (0645 MST) the shadow is just barely discernible. Five minutes later it is clearly defined and descends at a decelerating speed in response to a sun whose depression angle decreases by 0.9° every 5 minutes. At 0715, bottom right, the sun rises to cause Kitt Peak's shadow to emerge (white mark).

2.13 Alpenglow

High, snow-covered mountains glow yellowish orange when lit up by the low sun. Perhaps 30 minutes later after the sun has set, the [...]le which seems to brighten [...]ain glow is called the alpen-

[...] to find) little book *Die* [...]anation of alpenglow is [...]-orange light from the low [...] that is reflected from the [...]trast is striking against the blue sky (Figure 2.13A) and even more so against the earth shadow (Figure 2.13B). When the alpenglow consists of reflected sunlight, shadows can be seen on the mountain face. The twilight arch, on the other hand, is extended by many degrees along the horizon and shadows are not present or are greatly subdued.

Fig. 2.13A Mount Sefton in the Cook Range of New Zealand bathed in the yellow light of the low sun.

Fig. 2.13B Mount Sefton illuminated by the twilight arch to create a
shadowless alpenglow having lilac hues.

2.14 Purple light

When the sun is 4–6° below the horizon (about 30 minutes after sunset at mid-latitudes) a diffuse purplish glow 10–20° across may develop 45° above the sun (Figure 2.14A). This is the purple light. It may be visible for many months at a time and then fade away, not to be seen again for years. Photographing the purple light proves difficult because its range of brightness exceeds the capacity of color film.

The purple light is not fully understood but is almost certainly related to the scattering of light from stratospheric dust 20–25 kilometers up (Figure 2.14B). Direct reddened sunlight combines with indirect blue scattered light to produce a purplish hue[29]. Although the purple light's color is not hard to understand, its roughly circular or oval shape is problematical, especially in view of the stratified layering of stratospheric dust.

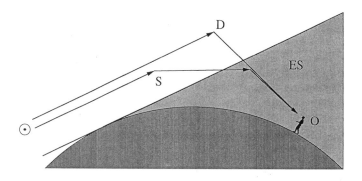

Fig. 2.14B Optics of the purple light. The purple light is a mixture of light scattered from the lower atmosphere S, which is in the earth shadow (ES) and sunlit dust (D) in the stratosphere. Because the amount of stratospheric dust is variable, so too is the occurrence of the purple light.

Fig. 2.14A The purple light is difficult to record photographically because of the sky's large range of brightness.

There is little debate about the purple light's intensification after major volcanic eruptions, but some controversy remains about whether stratospheric dust is needed for its occurrence at all. Some people claim that in the absence of dust there is no purple light, while Rozenberg[29] says that molecular scattering by stratospheric gas is enough to create a faint purple light. It seems reasonable to suppose that there is always some dust in the upper atmosphere, especially when meteoritic dust is considered.

2.15 'It's darkest just before dawn'

Is it really darkest just before dawn? We hear this so often that we might believe it. And any camper who has lain awake outside shivering on a cold night yearning for the warmth-giving sun to rise may well agree. But is it true?

The answer is no. Daylight and twilight both come from the sun. Twilight's brightness is a function of sun's distance below the horizon: the lower the sun the dimmer the twilight sky. The sun is farthest below the horizon halfway between sunset and sunrise and consequently midnight is the darkest hour.

Our belief in maximum darkness just before dawn may come from a secondary association of darkness with cold. The landscape starts to cool off late in the day and continues to cool all night long, warming only at sunrise. It truly is coldest just before dawn. But not darkest.

Is morning twilight different than evening twilight? Well, yes and no. All twilight phenomena are symmetric about midnight and, of course, occur in reverse order between sunset and sunrise. In this sense there should be no optical difference between dusk and dawn. Yet two factors are probably at work both to change our perception of the two events and actually to alter the atmosphere overnight.

The first is physiological and perhaps psychological. At sunset, our eyes are daylight adapted and may even be a bit weary from the day's toil. As the light fades, we cannot adapt as fast as the sky darkens. Some hues may be lost or perceived in a manner peculiar to sunset. At sunrise, however, the night's darkness has left us with very acute night vision and every faint, minor change in the sky's color is evident. We may also have just awakened from peaceful sleep, be well rested and eager to start the day.

The second reason has to do with the day's winds and urban activities that add dust and aerosols to the air. At sunset the sky is full of pollutants and wind-borne particles. During the night, winds die down, smog-producing urban activity eases and the atmosphere cleanses itself. The dawn is cleaner than any other time of day.

2.16 Twilight and volcanic activity

A sudden onset of thin, striated clouds at dusk, abnormally reddened sunsets, and a strengthening of the purple light are good evidence that somewhere in the world there has been a volcanic explosion. Why?

In our times it has been the eruptions of El Chichon in Mexico and Mt Pinatubo in the Philippines that lit up the twilight sky (Figure 2.16). El Chichon[30,31] went off like a cannon on April 4, 1982. Some 3–4 cubic kilometers of sulfur laden ash was propelled up to a height of 28 000 meters. While the heavier particles fell out quickly, the accompanying sulfur dioxide gas combined with water vapor to create a haze of sulfuric acid aerosol that remained aloft a long time.

The twilight–volcanic connection goes back to the eruption of Krakatoa on August 26, 1883. An Indonesian island ceased to exist on that date and tens of thousands perished in the ensuing tidal wave. Over the months following this cataclysmic event the northern hemisphere was treated to 'lurid' sunsets and a host of unusual atmospheric optical displays. Members of the Royal Society in London began inquiries into these happenings, and from their effort came *The Eruption of Krakatoa and Subsequent Phenomena*[18,19].

The relation between twilight intensification and volcanic eruptions is so well established that we can assume the latter if we observe the former. In the months following the eruptions of El Chichon and Pinatubo there were strong twilight effects and a noticeable increase in both the brightness of the aureole and atmospheric opacity.

Up-to-the-minute information on world-wide volcanism is available in the *Bulletin of the Global Volcanism Network*[32]. Also included is information on meteors, meteorites, fireballs, earthquakes, tsunamis, and any other sudden or unpredictable geophysical event.

Fig. 2.16 Striated sky at sunset during passage of the El Chichon ash cloud on May 7, 1982. The purple light was enhanced.

NORMAL ATMOSPHERIC REFRACTION

2.17 Atmospheric refraction

When the sun sets over water and its lower rim first touches the sea, the sun is actually below the horizon. We know this from timing the sun's motion across the sky. As it approaches the horizon it appears to slow down. It is, of course, moving at a steady rate because the earth's rotation speed is constant. The sun's apparent progress is slowed by atmospheric refraction, which bends its light and makes it appear higher in the sky than it actually is. For the most part, atmospheric refraction passes unnoticed. This is because we never get the chance to see the sky in the absence of the atmosphere.

Figure 2.17A shows how extraterrestrial light is refracted during its passage through the atmosphere. The amount of refraction depends on the index of refraction of air and the zenith distance z of the object.

The index of refraction n of air is a measure of how much it can bend light[33]. It is determined by the number and type of atoms or molecules per cubic centimeter. In the vacuum of space, n is equal to unity; it gradually increases on descent through the atmosphere where it reaches the grand value of about 1.000 294 1 in green light for pure air at sea level. Figure 2.17B shows the index of refraction of dry air as a function of wavelength. The presence of water vapor decreases n slightly. Above sea level, the amount of refraction is less because the air density is less. Normal atmospheric refraction always raises the image of an extraterrestrial object. We call this normal refraction because it is always present and is fairly predictable. Abnormal refraction occurs when there are unusual temperature gradients (hence density gradients) which produce unusual index of refraction profiles.

When an observer sees a star at an altitude a', the altitude of the star is actually slightly lower in the sky at altitude a. The amount of refraction R, equal to $(a - a')$, is zero at the zenith and increases toward the horizon (Figure 2.17C). In dry air the horizontal refraction is 39 arc-minutes for a sea-level observer. Since the angular diameter of the sun is only 30 arc minutes, when its lower edge appears to be just touching the apparent horizon, the entire sun is actually completely below that horizon.

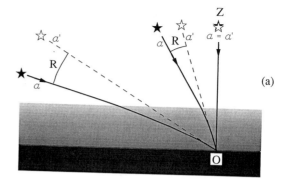
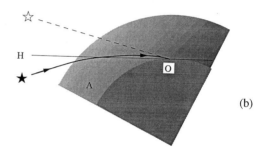

Fig. 2.17A Optics of normal atmospheric refraction. (a) When an observer sees a star at an altitude of a', the star is actually slightly lower in the sky at altitude a. The amount of refraction R, equal to $(a - a')$, is zero at the zenith and increases toward the horizon. (b) At the horizon, refraction causes objects that are below the geometrical horizon to appear.

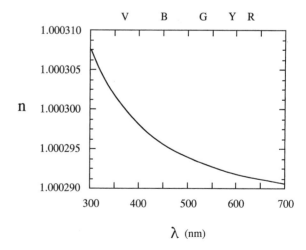

Fig. 2.17B Index of refraction n of air at sea level as a function of wavelength λ.

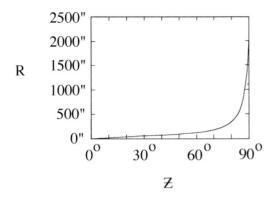

Fig. 2.17C Refraction R (in seconds of arc) as a function of zenith distance z.

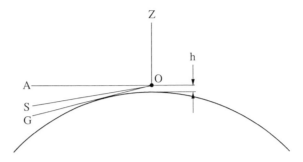

Fig. 2.18A Horizons. The observer O is located a distance h above ground. The astronomical horizon A is 90° from the zenith Z. The geometrical horizon G is where the observer's line-of-sight is tangent to the earth and is always more than 90° from the zenith. The sea-level horizon S is the location where the geometrical horizon is observed and is above it due to refraction. Our usual concept of horizon is the sea-level horizon. Unless the observer's elevation is large, the three horizons are very close to one another.

2.18 Horizons

The next time you are at the seashore, look out at the horizon. There it is, a hard line separating sea from sky. Clear, plain and simple, right? The horizon would seem to require no explanation. And yet, like most things under careful study, there is more to a horizon than meets the eye.

The *astronomical horizon* is an imaginary plane passing through the observer's eye and perpendicular to the zenith (Figure 2.18A). The *geometric horizon* is defined as the plane separating earth from sea. Owing to refraction, the visual location of the geometrical horizon is displaced upward from the astronomical horizon by a small amount to give us the *sea-level horizon*. There is an abrupt change in the amount of atmosphere in the line-of-sight as we go from looking just below the sea-level horizon to just above it. The sudden change in air mass, and thus refraction, at the geometrical horizon causes objects below that horizon to be refracted above it.

How far away is the geometrical horizon? The distance depends on the square root of the observer's altitude above sea level, provided the altitude is much smaller than the radius of the earth (Figure 2.18B). Because the distance to the horizon and thus optical depth both grow large with height, a hard, sharp horizon cannot be seen when the observer is higher than a kilometer or two, and often a good bit less when aerosols are present[34]. Also we can actually see 'over' the geometric horizon to some extent, increasing the true distance to the sea-level horizon by about 9%[35].

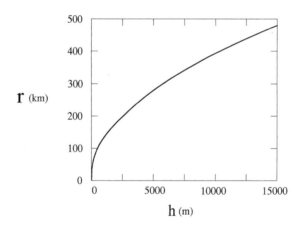

Fig. 2.18B Distance to horizon r in kilometers as a function of observer's elevation h in meters.

2.19 Flattening of the low sun and moon

Within a few degrees of the horizon, the sun (or full moon) no longer appears circular but is noticeably flattened (Figure 2.19A). In general, large amounts of flattening are correlated with dim, reddish suns.

The flattened sun near the horizon is caused by ordinary atmospheric refraction. The amount of refraction increases rapidly towards the horizon. Because the solar image has an appreciable

Fig. 2.19A The rising full moon as observed from Skylab by the astronauts. The flattening of its disk, by atmospheric refraction, is almost double that seen from the earth's surface because, from their vantage point, the astronauts can look through about twice the amount of air compared to the ground observer.

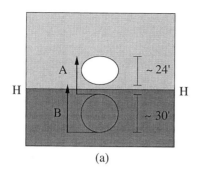

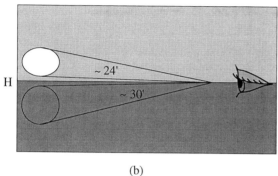

Fig. 2.19B Optics of the flattened sun. (a) The lower edge of the sun (B) is farther from the zenith than the upper edge (A) and therefore refraction raises it by a greater amount. (b) Typical values for refraction of the upper and lower edges are 24 and 30 arc minutes respectively.

2.20 Green flash

Provided we have a low horizon, preferably the sea, and a yellowish setting sun (not red), as the disk sinks from view the very last segment of the sinking sun will sometimes 'flash' green (Figure 2.20A). Much publicized, this is the green flash or 'green ray'[36-8].

Contrary to popular belief, green flashes are quite common, especially over water. Indeed, they happen to some degree every time the sun sets (or rises). The difficulty is in observing them. They only last for a second or two and they happen when the sun's light is rapidly fading. If we look at the solar disk directly, our eyes will be dazzled and there is a risk of missing the relatively faint ensuing green light. Observing green flashes at sunrise is difficult because the exact location on the horizon of the first rays is hard to predict. By the time our eyes turn to the bright spot, the flash is over. Because green flashes involve less than an arc minute (comparable to the resolution of the eye), the naked-eye observer can detect the green light but not resolve it. For this reason binoculars or a small telescope are helpful. A setting sun presents no danger to the eyes.

The duration of a green flash depends in part on the rate at which the sun sets. During summer in high latitudes, the sun approaches the horizon at a grazing angle and passes below the horizon so slowly that the green flash may last many seconds. On

angular size (about 30 arc minutes), light from the bottom of the sun passes through more air than does light from the top. It is therefore refracted upward by a greater amount than light from the top. This causes the lower part of the sun to be pushed upwards and closer to the upper part. The sun is not only elevated, it gets squashed (Figure 2.19B). Since the flattening depends on the altitude of the sun, the elevation of the observer, and how the temperature varies through the atmosphere, a wide range of flattening is possible. When the air is clean and the atmospheric temperature profile is normal, the amount of flattening at the horizon is about 20%.

When viewed from a high mountain the amount of flattening is greater than from sea level because of the added amount of air the light must pass through to reach the observer. From space[10], the flattening of the moon is about 40% (Figure 2.19A); the sun, being brighter, can be flattened more.

Fig. 2.20A Green flash above the horizon. The tiny segment above the distorted disk is a mock mirage[38]. (Photo by A. Young)

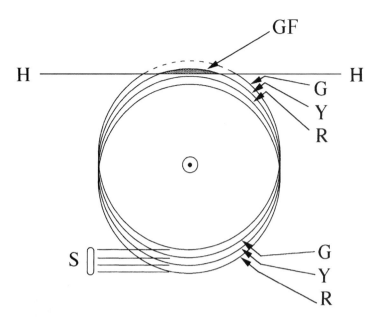

Fig. 2.20B Optics of the green flash. Dispersion in the atmosphere smears the setting sun by about an arc minute in the vertical direction, producing separate images S for each color. Because the sun is about 30 arc minutes across, the dispersion is only apparent at the top which is green (G) and the lower edge which is red (R). When all but the last bit of upper green rim has set, the green flash (GF) is visible.

the sky and the red lowest. As the total amount of dispersion is much less than the diameter of the sun the images overlap except for the extreme upper and lower edges. The lower part of the sun is red for the same reason that the upper part is green: dispersion (Figure 2.20C).

One might expect the upper edge to appear blue rather than green since refraction is greatest here (Figure 2.20C). Rayleigh scattering, especially from aerosols, tends to eliminate blue sunlight and atmospheric absorption by water vapor removes the yellow leaving primarily green and red.

O'Connell and Treusch[36] have described many interesting aspects of green flashes. Their photographs were obtained through a telescope sighting across the Mediterranean, but unaided visual observations were also made. In extremely clear air, shorter wavelengths of light may be present, leading to a blue flash. Chilean astronomers in La Serena on the coast report they frequently see a blue flash. In our experience, this really is a violet flash because, for reasons unknown, blue light is notably absent. Perhaps aerosols play a greater role than we expect or subtle aspects of human vision are involved. Our observations from Kitt Peak are visual using all reflective optics so that chromatic aberration is completely absent.

Recent work on green flashes shows them to be even more complicated than first thought[41]. There are two common types of green flashes, each associated with a characteristic shape of the setting sun (Figure 2.20A). In each case the green rim, which some have claimed is ordinarily too faint to be seen visually[42], is intensified and stretched vertically. The first is the 'inferior mirage flash' (Figure 2.20E) and is associated with a setting sun shaped like the Greek letter Ω (Figures 2.20F and 3.11C). As the sun approaches the horizon its inferior mirage appears below and seems to reach upwards and join the sun. As the sun sets, the last remaining part of the sun is a combination of real and miraged sun floating immediately above the horizon which turns green just before it disappears (Figure 2.20E). The second is called the 'mock mirage' flash and results from a temperature inversion in the low atmosphere. The mock mirage flash appears above the setting sun as a detached, green sliver (Figure 2.20A). Because there are so many possible temperature profiles in the atmosphere, the green flash can occur in many different forms and brightnesses.

certain days in polar regions when the sun never totally sets but rather skims along the northern (or southern) horizon with only its upper edge showing, the green flash may last many minutes. During Admiral Byrd's expedition to Little America in 1929, the green flash was observed on and off between the irregular ice floes for a period of 35 minutes[40].

The green flashes occur because of dispersion in the atmosphere (Figure 2.20B). Since the index of refraction, and thus the amount of refraction, depends on wavelength, all images passing through the atmosphere are dispersed vertically. Near the horizon the apparent sun is composed of a continuum of such images, each at its own wavelength and location. Refraction is greatest for the shorter wavelengths, so that the blue image of the sun is highest in

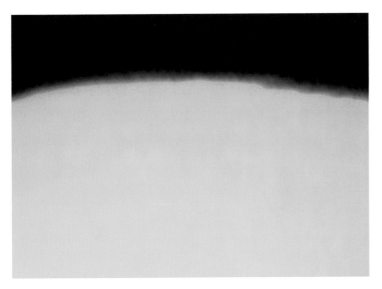

Fig. 2.20C The red and green rims of the lower and upper borders of the low solar disk. On the yellow image is a sunspot. The reason the direction of dispersion appears reversed in spots compared to the disk is that spots are dark while the sun is bright.

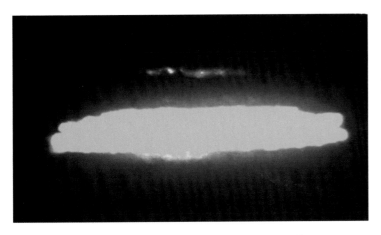

Fig. 2.20D Blue flash as seen from the South Pole Station, March 22, 2000. Taken through the eyepiece of a celestron C-8 about 48 hours after geometric sunset. The temperature was −63 °C and the display was visible intermittently for several hours as the sun skimmed the horizon. (Photo by R. Marks, deceased)

Fig. 2.20E Green flash (inferior mirage type) at the horizon. (Photo by K. Langford)

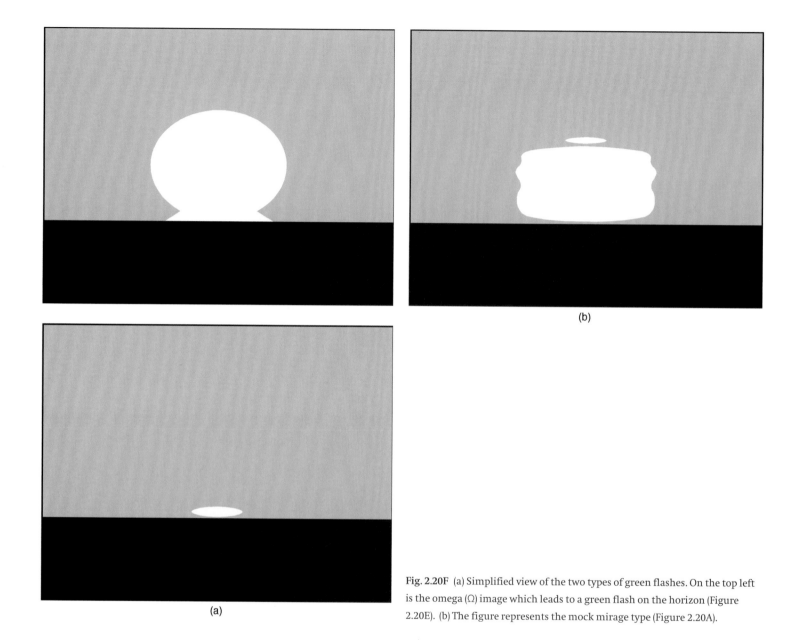

(b)

(a)

Fig. 2.20F (a) Simplified view of the two types of green flashes. On the top left is the omega (Ω) image which leads to a green flash on the horizon (Figure 2.20E). (b) The figure represents the mock mirage type (Figure 2.20A).

2.21 Twinkling

As everyone knows, stars twinkle. Twinkling is strongest low in the sky when it is clear and windy. The stars brighten and fade rapidly and near the horizon they may change color for an instant.

Twinkling is caused when starlight passing through the atmosphere encounters small, local variations in air density. Since the index of refraction of air is proportional to the density, the light finds itself continually being refracted and dispersed (Figure 2.21). Refraction from pockets of inhomogeneous air bends the rays momentarily away from our eyes. These 'seeing cells' act like a wind-blown procession of very weak prisms passing between us and the star. The star's brightness flickers irregularly many times a second, usually dimming momentarily as light once bound for our eye is refracted out of sight. Naturally, the amount of twinkling increases with the number of cells along the line-of-sight, so that twinkling is more vigorous near the horizon where the light passes through many air masses. At high elevations one sees less twinkling and in space, of course, there is none. Temperature fluctuations are primarily responsible for the density variations but they can also result from changes in the humidity.

Twinkling tends to be strongest on cold, clear, windy nights. Such evenings usually follow the passage of cold fronts and the post-frontal winds create a great deal of turbulence. The air is then more transparent by virtue of being dry and particle-free. This allows us to see stars very near the horizon where the air mass and number of density fluctuations are large[43]. Even distant street lights twinkle.

For an extended source like a mountain or the moon, the change in brightness of one tiny spot is not noticeable. For the same reason, planets display relatively little twinkling. Planets are not point sources like stars but rather they are extended objects. When light from part of their image is shifted by a few arc seconds out of the field of view, there is a good likelihood that light from another part will be refracted into the line-of-sight, negating any intensity change. Furthermore, the angular size of many seeing cells may be less than that subtended by a planet and thus any seeing cell would only affect a small part of the planet's light. In the extreme case of an extended object, twinkling is absent although shimmering or 'heat waves' may be present. Shadow bands are another visual manifestation of atmospheric seeing (Section 6.4).

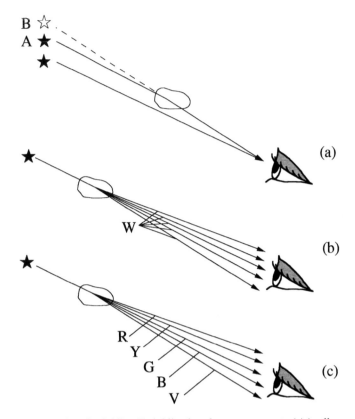

Fig. 2.21 Optics of twinkling. Twinkling has three components. (a) A cell passing between the star and observer will refract the starlight so that instead of coming from direction A it appears to come from direction B. (b) Refraction will also spread the beam of white starlight (W) so that not all of it reaches the observer's eye. When this happens the star momentarily fades. (c) Dispersion separates the colors of starlight so that an observer may see only part of the spectrum. When this happens he may momentarily see first red R, then green G then blue B.

UNUSUAL ATMOSPHERIC REFRACTION

2.22 Distortions of the low sun

Sometimes the setting sun is more than just flattened as described above, it is distorted and even fragmented to some degree (Figures 2.22A, 2.22B, 2.22C). Such irregular shapes result from refraction by localized, usually stratified density (i.e. index of refraction) variations in the lower atmosphere. O'Connell and Treusch[36] and Meinel and Meinel[22] present dozens of examples of these images. The same phenomena are reported for the low moon[44].

These distortions arise from variations in the index of refraction of air with height. This can happen when moist air overlays dry air or if there are unusual temperature gradients such as temperature inversions. These all work to create a non-uniform optical medium which can refract light rays in surprising ways. A sky with uniform color and brightness hides these atmospheric variations. Something bright and extended is needed to see them and the sun is the perfect source. The effects are most visible near the horizon[45] for two reasons. First, there is more air in the line-of-sight owing to the large air mass. Second, since the index of refraction variations usually occur in the vertical direction, a nearly horizontal line-of-sight is necessary to separate them.

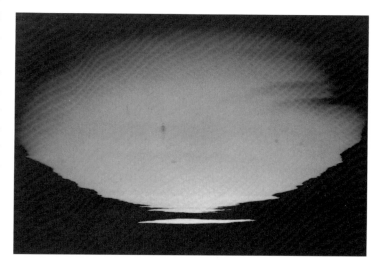

Fig. 2.22B Distortions of the low sun.

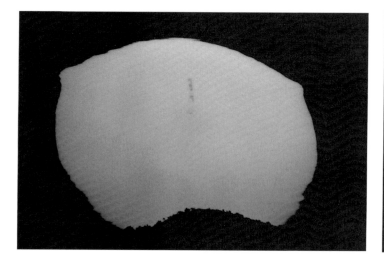

Fig. 2.22A Distortions of the low sun.

Fig. 2.22C Distortions of the low sun.

2.23 Mirages

A mirage is a refracted image of something that is normally not there. It is a result of unusual index of refraction profiles, particularly in the vertical. We emphasize 'unusual' because there is a continuum of atmospheric refraction conditions between what is normal and what is not. It is sometimes difficult to say when we have a mirage. For example, were it not for the fact that the flattened setting sun happens every day and is considered normal, we might say that it is due to a mirage. There are a wide range of possible atmospheric conditions and an equally wide range of mirages possible. Therefore we shall only describe those that are most commonly seen.

All mirages are confined to small angles, usually less than 1/2°, or about the size of the solar disk. Photographs of mirages, including those reproduced here, are generally taken with telephoto lenses which magnify these structures.

2.24 Inferior mirage

Everyone is familiar with the sight of shimmering water on a highway, and how that water miraculously 'evaporates' as we approach. Were we to look in the rear view mirror, we would find that the water reappears behind us. These days it is called the 'highway mirage' (Figure 2.24A). In times past it was more often referred to as a 'desert mirage', that notorious promised water that never materialized. The phenomenon is an inferior mirage, so named because we look below the horizon to see it (a sinking mirage). Inferior mirages can also be seen within a few meters of our eyes as we sight over a car roof on a hot day. Such mirages are found everywhere: near a toaster, a room heater, a sun warmed plank, etc.

The shimmering 'water' of a mirage is nothing more than blue skylight whose rays have been refracted upward by a thin layer of heated air just above the road (Figure 2.24B). The index of refraction of air is proportional to its density which, in turn, depends inversely on temperature. Near the highway surface, grazing rays from the sky in front of us pass down from higher cooler (hence denser) air to warmer rarefied air and back up again into our eyes. The amount of curvature of the ray depends on the temperature gradient: the larger the temperature gradient the greater the curvature. If a mountain rather than the sky is properly located, then we see mountain light refracted up to us from below the horizon. In this way, any object near the horizon can be seen in a mirage (Figure 2.24C). For a constant temperature gradient, the paths followed by light rays are parabolic.

Although we tend to think of mirages as summer phenomena, this is not always the case. The absolute temperature of the air is unimportant. What counts is how rapidly the temperature changes with height. Perfectly lovely highway mirages are seen in winter when cold air is heated by a sun-warmed asphalt. In fact, mirages are almost always present on any stretch of flat road.

If the temperature gradient is uniform, the miraged image is displaced from its original position but is not distorted. When the temperature gradient changes with distance above the surface (for example from 10° per meter near the road to 20° per meter higher up), the image may be upside down, a reversal that goes unnoticed if it is the sky that is miraged because there is no structure to tell up from down. But when something like a car is miraged, its upside down image is obvious (Figure 2.24D). More complex temperature gradients not only invert the image, they stretch them vertically or compress them. Such distortions are called towering and stooping, respectively. Different parts of the image may suffer both distortions and in very complex situations, multiple distorted images occur[46,47].

Fig. 2.24A Highway inferior mirage sequence of an approaching truck.

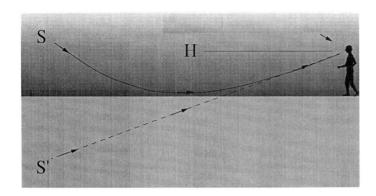

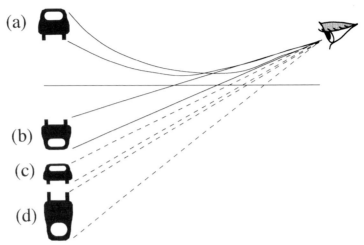

Figure 2.24B How inferior mirages are formed. Inferior mirages occur when a light ray S coming from above the horizon H is refracted by air whose index of refraction decreases rapidly toward the ground. The ray reaching the observer's eye is perceived to be coming from S′ below the horizon.

Fig. 2.24D Multiple images of car (a) that are (b) inverted (c) erect and stooped and (d) inverted and towering.

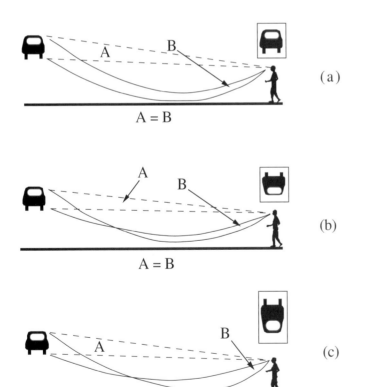

Fig. 2.24C When an extended object like a car is miraged, it can take one of many forms. (a) If the miraged angular height B is equal to its original angular height A and the upper and lower rays do not cross, the miraged image (shown in the box over the observer's head) is erect and the same angular size as the direct image. (b) If the rays cross the image is inverted. (c) If the rays cross and B exceeds A, the image is stretched vertically and said to be 'towering'. Towering images can be erect or inverted.

2.25 Superior mirage

On an ocean voyage or at the seashore one may be startled by an inverted image of a distant ship, steaming serenely above the direct image of the vessel. A plains dweller may occasionally see distant mountains where there are no mountains. Positioned above the horizon, these are instances of a superior mirage (Figure 2.25A).

When warm air overlies cold air, an atmospheric temperature inversion is said to exist; normally air temperature decreases with height (Figure 2.25B). Near horizontal light rays passing upward through such a warm layer are refracted downward and we see them above their non-refracted position as a superior mirage. This temperature inversion can be a consequence of the radiative cooling of the ground overnight, or a warm air mass overlying cold water. Light from a distant mountain which is refracted back downwards will then appear elevated, or looming. Often the mountains are also magnified in the vertical direction (towering). In Figure 2.25A we look across a desert plain at dawn to find a towering image of distant hills topped by its inverted replica. Superior mirages are even capable of bringing into view objects below the horizon. It is said that Greenlanders occasionally caught sight of North America via superior mirages long before Europeans 'discovered' the land.

As with inferior mirages, superior mirages can take on a variety of properties depending on the temperature profile of the air. In addition to simple towering and inversion, three other types of superior mirages are noteworthy: (1) the hillingar effect, (2) the hafgerdingar effect, and (3) the Novaya Zemlya effect[48,49].

The hillingar effect causes the earth and horizon to be refracted in such a way as to appear flat or slightly concave upwards. It is caused by a mild, uniform and wide-spread temperature inversion.

The hafgerdingar effect causes distant objects and the horizon to grow spikes and shoots ('castles in the air'). These complex mirages are traditionally referred to as Fata Morgana (the Morgan Fairy), after an Italian expression stemming from 15th century descriptions of Mediterranean mirages that were then inexplicable. The spires are caused by strong, non-uniform temperature inversions.

Another fascinating mirage is the Novaya Zemlya in which the wildly distorted, multiple-imaged sun is seen above the horizon even though geometrically, the sun has set. In the Novaya Zemlya, light becomes trapped at a complex temperature inversion and is ducted over the horizon in a series of oscillating trajectories, traveling first up, then down, then up again. The trapped rays carry the sun's light for hundreds of kilometers over the horizon. Yet another superior mirage, peculiar to polar regions is the Fata Bromosa (fairy fog). This appears as a featureless horizontal white band over water or in arctic regions, due to mirages of snow fields.

Mirages can also occur at night. Before dawn while driving through a high inter-mountain valley in Colorado, we saw a string of vertical lights which flickered and varied in spacing. A comparison with the full moon allowed us to gauge that the display subtended less than $1/2°$. Eventually we passed the car whose headlights were the probable source of this superior mirage.

Finally, the mirage need not take place close to the ground. Figure 2.25C shows such a mirage at about 1000 meters above the ground.

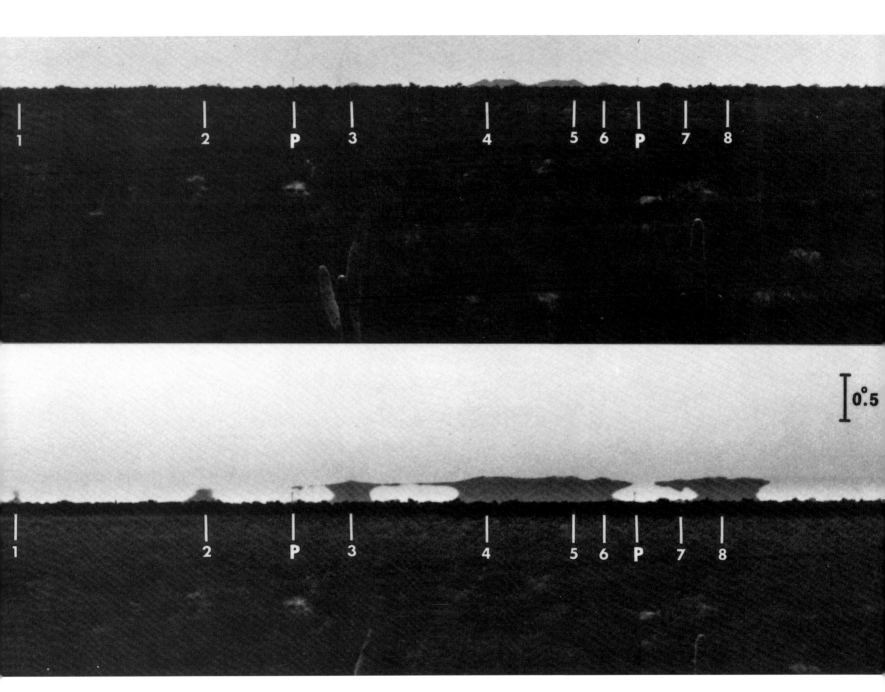

Fig. 2.25A Desert superior mirage showing 'direct', inverted, and erect images as an example of over-the-horizon viewing. Notice that the power poles P, P in the intermediate distance appear undistorted. The furthest miraging features (1,2,8) are 58 kilometers away and over the horizon. (Las Guijas Mountains, Altar Valley, Arizona)

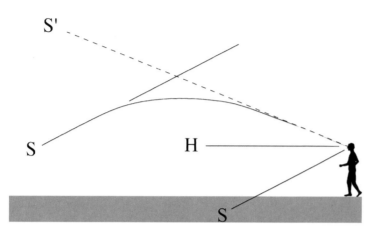

Fig. 2.25B Optical diagram for the superior mirage. A superior mirage occurs when light from an object S below the horizon H is refracted downward so that it appears to be at location S' above the horizon.

Fig. 2.25C Mirage at high elevation. (Photo by A. K. Pierce)

Fig. 2.25D Top, miraging water wave crests. A common sight through binoculars in which wave extremities are exaggerated by looming. Bottom, Fata Morgana at work on Saguaro cactus.

1 deg

2.26 Lateral mirage

If you look carefully along a sunlit east–west wall you may see what looks like a reflection of the landscape just beyond the wall (Figure 2.26A). This is called a lateral mirage[10] and is nothing more than an inferior mirage created by a lateral temperature gradient where the wall is hotter than the air a few centimeters away. A north–south west-facing wall can produce a superior mirage when the wall is cooler than the air.

Figure 2.26B shows a stick leaning against a wall and its mirage. The mirage contains three portions which make the end of the stick appear jagged and twisted.

By varying the distance of our eye from the wall we may explore the entire gamut of inferior mirages generated by the existing temperature profile.

Fig. 2.26A Lateral mirages of a person next to a wall. At least two mirages can be seen.

Fig. 2.26B Lateral mirage of a stick leaning against a solar heated wall. Three mirage components are visible.

2.27 Airglow

The entire upper atmosphere continuously emits a feeble light called airglow. It adds a uniform featureless glow to the night sky and is the principle source of light in the night sky. Indeed, on a dark clear night a tree or building or even your hand will appear totally black silhouetted against the night sky's light. Much of that light is airglow[50]. On rare occasions patches of diffuse 'moving bands' have been attributed to the airglow[51]. The airglow is strongest at mid and lower latitudes.

Airglow is photochemical luminescence from atoms and molecules in the ionosphere at elevations of 60–300 kilometers with a concentration around 110 kilometers. The emission comes primarily from molecular oxygen and nitrogen, the hydroxyl (OH) molecule and from atomic sodium and oxygen. Solar ultraviolet radiation does one of two things: it dissociates the molecules, or ionizes or excites the atoms or molecules. When the particles recombine or become de-excited they emit light. The light is the airglow.

2.28 Aurora borealis (northern lights)

Aurorae are multicolored, diffuse, slowly moving lights seen in high latitudes against the clear dark night sky. The variety of color, shape, and movement of aurorae seem limitless. They are often greenish-yellow, though colors from violet to red have been reported. The shapes are classified as auroral arcs (smooth curves), bands or 'curtains' showing folds with fairly sharp lower boundaries and diffuse upper edges, homogeneous patches that resemble ordinary clouds, veil aurorae (large scale nearly featureless glows covering a large part of the sky), and long, nearly straight rays. Although their movement is normally slow and easily followed with the eye, they may pulsate, flicker, or appear as moving waves. The proper name is aurora borealis which means northern lights (aurora australis in the Southern Hemisphere).

Aurorae occur in two forms, diffuse and discrete. Diffuse aurorae are present all the time and generally pass unnoticed except as a general brightness of the whole night sky. In this sense they are similar to the airglow. Discrete aurorae are most spectacular during and a few years either side of sunspot maximum, especially around the times of the autumnal and vernal equinoxes. They are at their best along the auroral oval where they can be seen every clear night[52,53].

Aurorae are caused by high speed particles, principally electrons, from solar active regions and flares that stream outward from the sun. They reach the earth's magnetic field and are channeled along its field lines toward the magnetic poles where they collide with atmospheric air particles. Reactions take place including molecular disassociation, excitation, and ionization. When the particles recombine or become de-excited, they emit auroral light.

Being atomic emissions, auroral light consists of narrow band line emission just as the airglow does. The common yellow-green light is due to atomic oxygen emission at 557.7 nanometers. Red may be from either atomic oxygen or molecular nitrogen. Aurorae seen from low latitudes are almost always a shapeless red glow.

The altitude and form of the aurora depend on the depth to which the solar particles penetrate. A typical band may be thousands of kilometers long, a few hundred kilometers high but only a few hundred meters thick.

Owing to the increasing strength of the earth's magnetic field toward the ground, solar electrons that do not strike air particles reverse direction and spiral back and forth between north and south poles, creating two mirror image aurorae, one in each hemisphere.

Aurorae stand alone for mouth-gaping, awe-inspiring, spellbinding majesty. Their silent play of eerie color is surely one of Mother Nature's grandest spectacles. With no reservation whatsoever we can recommend hopping on the first plane to the polar regions following a major solar flare if for no reason other than to witness these haunting dancing lights.

Fig. 2.28A Multiple auroral rayed bands with the constellation of Orion reflected in the Yukon River, Alaska. Note the upper part of the right band is slightly pinkish, probably caused by 630 nanometer line emission of atomic oxygen. The yellowish lower bound on the left likely indicates deeply penetrating electrons. (Photo by R. Hutchinson)

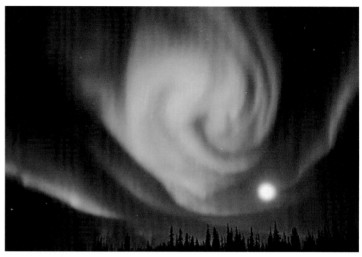

Fig. 2.28C Homogeneous bands, described as a serpent devouring the moon by R. Hutchinson, photographer.

Fig. 2.28B Curled bands. In this case the yellowish lower bound may be simply a distance effect from atmospheric absorption. (Photo by R. Hutchinson)

Fig. 2.28D Homogeneous and rayed bands. Again the yellowish lower bound at the bottom may be a consequence of distance. (Photo by R. Hutchinson)

Fig. 2.28E A series of homogeneous bands plus clouds. (Photo by R. Hutchinson)

Fig. 2.28F Auroral bands with twilight arch from Fairbanks, Alaska. (Photo by J. Curtis)

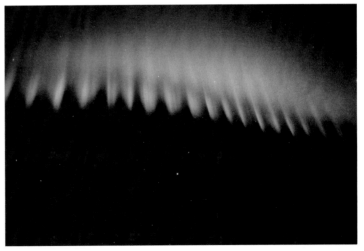

Fig. 2.28H Rayed band. (British Antarctic Survey)

Fig. 2.28G Auroral bands with rays on 6 September, 1996. (Photo by J. Curtis)

Fig. 2.28I Auroral storm on April 7, 2000 over Europe produced this red corona and rays. It was visible in North America as far south as Mexico. (Photo by Jan Safer, Brno Observatory, Czechoslovakia)

HOW TO PHOTOGRAPH THE AURORA

The aurora is always present within the *auroral oval*, a nearly circular narrow band centered on the earth's magnetic pole. Barrow, Alaska, and Churchill, on Hudson Bay, Canada, are two favorable sites, see Table 2.28. The latter, although at a lower latitude than Barrow, is favored because of the eccentric position of the oval. (In the southern hemisphere there are no good sites outside of Antarctica.) However, when the sun becomes active the width of the oval increases and the chances of observations from lower latitudes increases to 10% or better near the Canada–US border, northern British Isles, Scandinavia, and northern Asia. Other considerations include the need for clear skies, and a less than gibbous moon. Any aurora sightings in the summer season are ruled out by nocturnal twilight.

Dick Hutchinson, who lives in Circle, Alaska, offers the following advice on photography. Like ourselves he prefers Ektachrome 100S because of low reciprocity failure. Exposure is 10 to 25 seconds at f/1.4. He points out that a short focal length lens means the star trails are short and therefore less distinctive.

Table 2.28. *Percentage of nights on which aurora is expected.*

City	Percentage
Barrow, Alaska	100
Churchill, Canada	100
Fairbanks, Alaska	90
Tromsö, Norway	90
Kiruna, Sweden	80
Anchorage, Alaska	30
Winnipeg, Canada	20
Calgary, Canada	18
Oslo, Norway	10
Montreal, Canada	10
Bangor, Maine	9
Edinburgh, Scotland	8
New York, New York	4
Moscow, Russia	3
Denver, Colorado	3
Melbourne, Australia	3
Sydney, Australia	1
Capetown, South Africa	0.5
Los Angeles, California	0.5
Rome, Italy	0.1
Mexico City, Mexico	0.05
Buenos Aires, Argentina	0.01
Tokyo, Japan	0.01

After Eather[52].

References

1 Shackleton, E., 1920, *South!*, Macmillan Co., NY.

2 Gedzelman, S.D., 1980, *The Science and Wonders of the Atmosphere*, John Wiley and Sons, NY.

3 Henderson, S.T., 1970, *Daylight and Its Spectrum*, American Elsevier, NY.

4 Rayleigh (Lord Rayleigh, J.W. Strutt), 1871a, 'On the light from the sky, its polarization and color', *Philosphical Magazine*, **41**, 107–20.

5 Rayleigh (Lord Rayleigh, J.W. Strutt), 1871b, 'On the scattering of light by small particles', *Philosophical Magazine*, **41**, 447–54.

6 Rayleigh (Lord Rayleigh, J.W. Strutt), 1899 'On the transmission of light through an atmosphere containing small particles and the origin of the blue sky' *Philosophical Magazine*, **47**, 375.

7 Young, A.T., 1981, 'Rayleigh scattering', *Applied Optics*, **20**, No. 4, 533–5.

8 Young, A.T., 1982, 'Rayleigh scattering', *Physics Today*, **35**, 42.

9 Hess, P., 1939, *Gerlands Beitr. z. Geophys.*, **49**, 71–96.

10 Garriott, O.K., 1979, 'Visual observations from space', *Journal of the Optical Society of America*, **69**, 1064.

11 Lee, R., 1994, 'Horizon brightness revisited: Measurements and a model of clear-sky radiances', *Applied Optics* **33**, 4620.

12 Minnaert, M.J.G., 1954, *The Nature of Light and Colour in the Open Air*, Dover Publications, NY.

13 McCartney, E.J., 1976, *Optics of the Atmosphere*, John Wiley and Sons, NY.

14 Bohren, C.F., 1987, 'Multiple scattering of light and some of its observable consequences', *American Journal of Physics*, **55**(6), 524–33.

15 Coulson, K.L., 1968, 'Effect of surface reflection on the angular and spectral distribution of skylight', *Journal of Atmosphere Science*, **35**, 759–70.

16 Fraser, R.S., 1968, 'Atmospheric neutral points over water', *Journal of the Optical Society of America*, **58**, 1029.

17 Lynch, D.K., 1991, 'On step brightness changes of distant mountain ridges and their perception', *Applied Optics*, **30**, No.24 (Aug 20) 3508–13.

18 Lamb, H.H., 1977, 'Volcanic dust in the atmosphere; with a chronology and assessment of its meteorological significance', *Philosophical Transactions Royal Society*, **A266**, 425.

19 Symons, G.J. (Editor), 1888, *The Eruption of Krakatoa and Subsequent Phenomena*, Royal Society, Harrison and Sons, London.

20 Green, H.L., Lane, W.R. and Hartley, H., 1964, *Particulate Clouds: Dusts, Smokes, and Mists*, Van Nostrand, New Jersey.

21 Sassen, K., Peter, T., Luo, B.P. and Crutzen, P.J., 1994 'Volcanic Bishop's Ring: evidence for a sulfuric acid tetrahydrate particle aureole', *Applied Optics*, **33**, 4602–6.

22 Meinel, A.B. and Meinel, M., 1983, *Sunsets, Twilights and Evening Skies*, Cambridge University Press, Cambridge.

23 Grunner, P. and Kleinert, H., 1927, *Die Dammerungserscheinungen*, Verlag von Henri Grand, Hamberg.

24 Heim, A., 1912, *Luft-farben*, Hofer and Co., Zurich.

25 Neuberger, H., 1951, 'General meteorological optics', in *Compendium of Meteorology*, American Meteorological Society, Boston.

26 Neuberger, H., 1940, 'Some remarks on the problem of the dark segment', *Bulletin of the American Meterological Society*, **21**, 333–5.

27 Hulburt, E.O., 1953, 'Explanation of the brightness and color of the sky, particularly the twilight sky', *Journal of the Optical Society of America*, **43**, 113–18.

28 Adams, C.N., Plass, G.N., and Kattawar, G,W., 1974, 'The influence of ozone and aerosols on the brightness and color of the twilight sky', *Journal of Atmosphere Science*, **31**, 1662–74.

29 Rozenberg, G.V., 1966, *Twilight, A Study in Atmospheric Optics*, Plenum Press, NY

30 Mitchell, J.M., 1982, 'El Chichon, weather maker of the century?', *Weatherwise*, **35**, 252.

31 Livingston, W. and Lockwood, G.W., 1983, 'Volcanic ash over Arizona in the spring of 1982: astronomical observations', *Science*, **220**, 300.

32 *Bulletin of the Global Volcanism Network*, Smithsonian Institution, National Museum of Natural History, Washington, DC 20560.

33 Penndorf, R., 1957, 'Tables of the refractive index for standard air and the Rayleigh scattering coefficients for the spectral region between 0.2 and 29 μm and their applications to atmospheric optics', *Journal of the Optical Society of America*, **47**, 176.

34 Bohren, C. and Fraser, A., 1986, 'At what altitude does the horizon cease to be visible?', *American Journal of Physics*, **54**, 222.

35 French, A.P., 1982, 'How far away is the horizon?', *American Journal of Physics*, **50**, 795.

36 O'Connell, D.J.K. and Treusch, C., 1958, *The Green Flash and Other Low Sun Phenomena*, North Holland Pub. Co., Amsterdam.

37 Shaw, G.E., 1973, 'Observations and theoretical reconstructions of the green flash', *Pure and Applied Geophysics*, **102**, 223–35.

38 Bohren, C.F., 1982, 'The Green Flash', *Weatherwise*, **35**, 271–4

39 Young, A.T., Kattawar, G.W., and Parviainen, P. 1997, 'Sunset Science. I. The Mock Mirage', *Applied Optics*, **36**, 2689.

40 Haines, W.C., 1931, 'The green flash observed October 16, 1929 at Little America by members of the Byrd Antarctic Expedition', *Mon. Weather Review*, **59**, 117.

41 Young, A.T., 1999, 'Green flashes and mirages', *Optics and Photonics News*, **10**, 31.

42 Dietze, G., 1955, Die Sichtbarkeit des grünen Strahls, *Zeitschrift Für Meteorologie*, **9**, 169.

43 Keller, G., 1955, 'Relation between the structure of stellar shadow band patterns and stellar scintillations', *Journal of the Optical Society of America*, **45**, 845.

44 Dougherty, T.R., 1988, '1988 photo contest winners', *Weatherwise*, **41**, 222.

45 Ives, R., 1945, 'Sunset shadow bands', *Journal of the Optical Society of America*, **35**, 736.

46 Fraser, A.B. and Mach, W.H., 1976, 'Mirages', *Scientific American*, **234**, 102 (Jan).

47 Tape, W., 1985, 'The topology of mirages', *Scientific American*, **252**, 120.

48 Lehn, W., 1979, 'The Novaya Zemlya effect: An arctic mirage', *Journal of the Optical Society of America*, **69**, 776–81.

49 Lehn, W. and Schroeder, I., 1981, 'The Norse mermen as an optical phenomenon', *Nature*, **289**, 362–6.

50 Roach, F.E. and Gordon, J.L., 1973, *The Light of the Night Sky*, D. Reidel Publishing Co., Dordrecht.

51 Peterson, A.W., 1979, 'Airglow events visible to the naked eye', *Applied Optics*, **18**, 3390–3.

52 Eather, R.H., 1980, *Majestic Lights: The Aurora in Science, History, and the Arts*, American Geophysical Union, Washington.

53 Akasofu, S., 1989, 'The dynamic aurorae', *Scientific American*, **261**, May, 90–7.

3 Water and light

We dropped anchor for a swim at the Padre Island National Seashore. It was just before midnight but far from dark. Leaping overboard the water reacted with tremulous green surges. Every breaking wave pulsed with light and each of us was outlined in swirling turquoise. Trails of fish could be seen darting here and there, unseen except for their glowing subsurface wakes. After this dazzling light show, we sailed southward and watched our boat's luminous green tail. Out of nowhere , a pack of dolphins came bounding up. Their exuberant leaps from the water were preceded by a lime-green explosion at the surface as the creatures, now outlined like ghosts from below, cruised through the black night. The phosphorescent sea then carried us into sunrise.

Adapted from the authors' observing books

3.1 Light from water

Light from water comes from three places; the top, middle and bottom (Figure 3.1). This light may be:

(a) reflected from the water's surface,
(b) refracted through the top and scattered from the volume of water itself, then refracted though the surface, or
(c) refracted through the top, transmitted by the water before being reflected from the bottom, then refracted a second time through the air-water interface.

Together all three components make up the blend of light that we see. Under certain circumstances one may dominate the others. For example, in shallow water, light from the bottom is most important. In deep clear water where very little light comes from below, surface reflections may be the brightest at low angles to the horizon. In muddy water the light is scattered primarily from the suspended sediments near the surface.

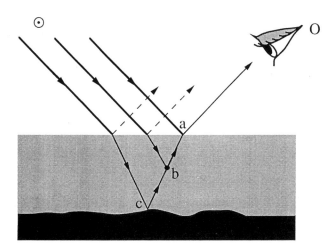

Fig. 3.1 Light from water. The light we see comes from three places. It is (a) reflected from the surface, (b) scattered from within the volume of water itself, and (c) reflected from the bottom.

3.2 Color of pure water

In trying to decide what color water is, what are we to make of the deep blue ocean with its white foam, its shift to gray on an overcast day, the turquoise of a Caribbean lagoon, red-orange seascapes at sunset and strangest of all, a perfectly clear glass of water? Does water have its own true color? If so, what is it and why are certain bodies of water so changeable[1,2]?

If sufficiently thick, water is blue, like a piece of blue glass[3]. And like glass or any relatively transparent substance, the thicker it is, the deeper the color. A raindrop is crystal clear. A bathtub full of water is faintly blue. A swimming pool is noticeably blue at its deepest end and the ocean is very blue indeed[4-6]. Water transmits light of every visible color, though it is clearest for shorter wavelengths (Figure 3.2A). It absorbs most at longer wavelengths in the orange and red part of the spectrum, and to a lesser extent in the blue and violet. Its peak transparency is in the blue-green near 480 nanometers. Water's tint is most apparent when viewed from under its surface.

From above, deep water's color is usually masked by reflected light from the sky and landscape. The ocean's color near the horizon comes primarily from blue reflected skylight. If the sky is overcast, water appears an uninteresting gray, a mixture of faint subsurface blue light and reflected white light from the clouds. Looking towards a sunset the water's red-orange color simply mimics the twilight sky. Because the reflectivity of water increases as the line-of-sight becomes more horizontal, the percentage of skylight reflected from its surface increases toward the horizon (Figure 3.2B). From a distance, airlight may interpose its own blue veil on the water.

In clear, shallow water, the bottom is well lit and it contributes more to the scene than does the invisible bottom of a deep ocean. Being relatively transparent, very shallow water adds virtually no blue light of its own. A vertical line-of-sight reveals only the bottom and any light patterns caused by surface waves. A more horizontal aspect shows a mixture of skylight and bottom-reflected light.

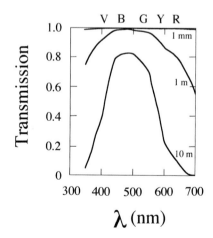

Fig. 3.2A Transmission T of water for depths of 1 millimeter, 1 meter and 10 meters.

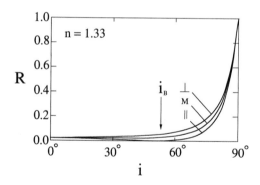

Fig. 3.2B Reflectivity R of water is the fraction of light reflected as a function of angle of incidence i. The amount of light reflected also depends on the plane of polarization measured perpendicular ($\perp$) and parallel ($\parallel$) to the plane of incidence of the light beam. At the Brewster angle ($i_B = \tan^{-1}n$) where the reflected and transmitted beams are 90° apart, no light is reflected by the parallel component (Section 3.8). The average reflectivity of water (M) is between $R_\parallel$ and $R_\perp$.

3.3 Color from suspended particles

Even the purest mountain lake contains tiny particles which scatter light. Silt, pollen, algae, bacteria, and wind-blown dirt all find their way into the water and affect its appearance.

When the reddish-black water of the Negro river meets the yellow ocher flow from the Amazon near Manaus, Brazil, the contrast is striking. These lazy rivers do not mix immediately (Figure 3.3A) but instead meander cheek-to-cheek, black to the north and yellow to the south. Their distinctive coloration is from various algae which thrive on the very different nutrients in the two rivers. A similar situation is found at the junction of the Green and Colorado Rivers in Utah (Figure 3.3B). Here the rivers' colors show a striking green–brown contrast due to the diverse mineral content of their suspended sediments. Ponds are prone to algae growth which may turn them green, red or brown. Fine, suspended silt renders lakes fed by glaciers a turquoise color (Figure 3.3C). Glacier-fed streams are characteristically milky blue and hot springs may be yellow due to suspended sulfur (Figure 3.3D).

Fig. 3.3A Confluence of the Amazon and Negro rivers illustrates how these rivers' waters are heavily colored by their algae content. Although the Negro's water appears black, a swim in its piranha-infested waters shows that it is actually blood red.

Fig. 3.3B Confluence of the Green and Colorado rivers displays the difference in the silt content of the two streams.

Fig. 3.3C Turquoise waters of Lake Tekapo, New Zealand. Color arises from
the inherently blue water plus a whitish scattering from fine glacial silt
carried by the water.

Fig. 3.3D Yellow water due to suspended sulfur particles. (Waiotapu, New Zealand)

3.4 Red tide and phosphorescent seas

During warm days the ocean sometimes develops rust-colored patches. Known as the red tide, this color comes from plankton bloom, a sudden population explosion. These blooms are promoted by an excess of nutrients and elevated water temperatures, both of which are found near shore. Plankton blooms sometimes happen in freshwater as well.

At night warm ocean waters often show blue-green sparkles, especially when the water is disturbed. Surf, fish, and boat wakes are good places to see them. These swirling lights make scuba diving at night an unforgettable experience. Each momentary flash is due to a single bioluminescent plankton appropriately named *noctiluca scintillans* (night lights). When the plankton population is large, the individual lights blend together to form a continuous glow. Though commonly known as phosphorescent water, the term is a misnomer since the light is produced chemically in short bursts.

Fig. 3.5A Mirror image produced by the still waters of Helen's Lake near Muir Pass in California's High Sierra. Distant objects seem to mirror perfectly but the foreground is a little darker because of reduced reflectivity in the water at steeper line-of-sight angles. Closer to the observer, perspective hides a few details of the upper part of the rocky island at the right.

3.5 Myth of a mirror image

What a sight: mountain peaks with their snow fields reaching into a blue sky, all perfectly reflected in the still waters of a mountain lake (Figure 3.5A). But look closer. The mirror image is not perfect.

First, the reflected image is not quite as bright as the original. At the horizon the two views have equal brightness but near the observer the reflected image is fainter. This is because the reflectivity of water is not 100% except at an angle of incidence of 90° (Figure 3.5B).

Second, subsurface objects become visible near the observer. As we look down more directly into the water (corresponding to smaller angles of incidence and shorter lines-of-sight) the reflected image grows fainter and transmitted bottom light

brighter. The reflection coefficient of water decreases with decreasing incidence angles causing the reflected image to fade. At the same time, the transmission coefficient for light reaching the surface from below increases. This allows more subsurface light to reach the observer. For vertical viewing only about 3% of the external light is reflected and most of the light comes from below.

Third, the vantage points are different. As Figure 3.5C shows, the landscape in the mirror image is always viewed from a slightly different direction than the direct image. For nearby objects viewed with a small angle of incidence, the difference between vantage points noticeably foreshortens the mirror image. For distant objects seen nearly horizontally, this difference is small and the mirror image is faithful to the original.

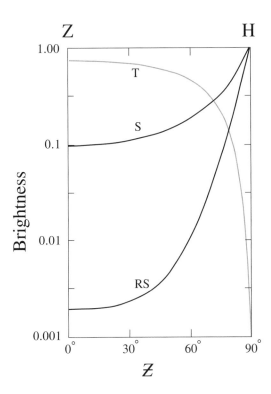

Fig. 3.5B The relative brightness of the sky (S) as a function of zenith distance at a wavelength of 500 nanometers (green). The brightness of the reflected sky (RS) in flat water is also shown. At the horizon (H) both the sky and reflected sky have the same brightness. The transmission of water (T) upward through the air–water interface is also shown. At the horizon there is no transmission of light from below while at the zenith it is almost 100%.

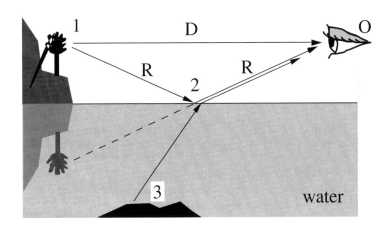

Fig. 3.5C When the observer (O) views reflected images of nearby objects he finds three differences between the direct (D) and reflected (R) image.
(a) perspective: the direct ray (1→D–O) leaves the object at a different angle than does the reflected ray (1→2→O); (b) brightness: some of the light from the object (1→2) is reflected downward into the water and lost, therefore the ray (2→O) is fainter than the direct ray (1→D→O); and (c) image contamination: light from objects below the surface (3→2→O) merges with the reflected ray (2→O).

3.6 Refraction through the air–water interface

Refraction, or bending, of light passing through an air–water interface causes the apparent location of objects to be displaced in the observer's vertical plane (the plane containing the observer and object) (Figure 3.6A). The exact amount of deviation, D, depends on the angle that the line-of-sight makes with the surface. In Figure 3.6B the spear fisherman looks in the direction OF′ and sees the fish. The fish, however, is actually in the direction OF. Similarly, the fish sees the observer in the direction FO′ even though the true direction is FO.

To catch the fish a spear fisherman must abandon his visual cue that tells him the fish is at F′ and aim instead lower at F. Of course he has no direct way of knowing this direction, so experience guides him. Conversely, the famous archer fish of southeastern Asia that knock insects from low branches must aim low, at O, even though it sees its next meal at O′.

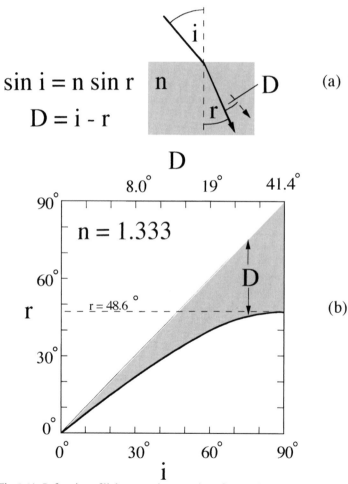

(a)

(b)

Fig. 3.6A Refraction of light at an air–water interface. An incident beam making an angle i with the normal is refracted through the interface where it is bent and makes an angle r with the normal. The deviation of the ray D ($= i-r$) increases from zero at normal incidence to 41.4° at grazing incidence.

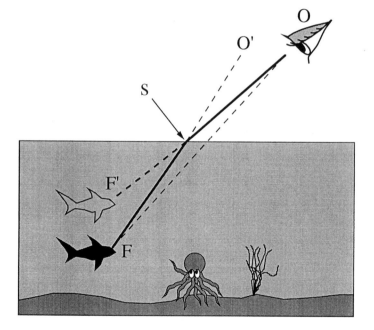

Fig. 3.6B How refraction displaces the apparent position of underwater objects. The observer at O sees the fish in the direction OF′, even though the direction to the fish is actually OF. From under water, the fish sees the observer at O′, even though the observer is actually at O. The angle OSO′ is the angle D by which the light is deviated.

3.7 Optical manhole

From under water the entire celestial hemisphere is compressed into a circle with a radius of only 48.6° (diameter 97.2°) (Figure 3.7A). Called the optical manhole[7], the effect occurs because light from the horizon (angle of incidence equal to 90°) is refracted downward at an angle of 48.6°. Rays approaching the surface from below, at angles greater than 48.6°, are totally reflected back downwards and never penetrate the surface into air. Since these rays originate below the surface, they are faint and provide the darkness that surrounds the manhole (Figure 3.7B).

Fig. 3.7A The optical manhole. From under water, the entire celestial hemisphere is compressed into a circle only 97.2° across. The dark boundary defining the edges of the manhole is not sharp due to surface waves. The rays are analogous to the crepuscular type seen in hazy air, Section 1.9. (Photo by D. Granger)

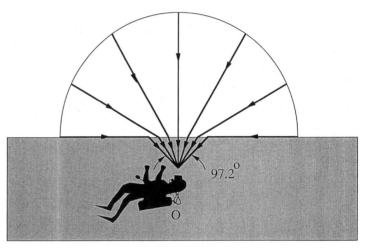

Fig. 3.7B The optical manhole. Light from the horizon (angle of incidence = 90°) is refracted downward at an angle of 48.6°. This compresses the sky into a circle with a diameter of 97.2° instead of its usual 180°.

3.8 Polarization and fish-locating goggles

Light reflected from the surface of water is partially polarized as Figure 3.2B shows. At the Brewster angle of incidence i_B ($\tan^{-1}n$ or about 53.1° for an index of refraction n of 1.33) the component of light perpendicular to the surface is completely transmitted, resulting in 100% polarization for the reflected beam. This reflected plane-polarized beam can then be totally eliminated with polarized sunglasses leaving only light coming from below the surface. The ability to see directly into water without any distracting surface reflections is the principle on which 'fish-locating' goggles work. Of course, they are only perfect when viewing objects at the Brewster angle.

3.9 Visibility of waves and the horizon

Have you ever seen a boat floating in the sky? It seems to hang somewhere between sea and sky (Figure 3.9A). But where is the horizon?

The horizon is the apparent boundary between two different sources of light: the sea and sky. To see the horizon, these two sources must contrast with each other, differing either in color, brightness or both. If they do not, then there is no visible distinction between them. This can happen when the water is completely smooth and flat. Light from water at the horizon is skylight that has been reflected from the surface with an incidence angle of about 90°. At this angle water's reflectivity is 100% and light reflected from it has the same color and brightness as the skylight. The lack of wave texture renders the sea featureless, like the sky. Consequently there is no contrast with the adjacent sky above and the horizon disappears.

The horizon over water is only visible when water waves disturb the surface and darken the water. Even the gentlest of swells is enough to reveal it. As the sea grows rougher, it darkens and thereby increases its contrast with the air.

Why do waves darken water? Our nearly horizontal line-of-sight strikes only those parts of the waves that face towards us (Figure 3.9B). The portions of the waves facing away from us are not visible. Even though the average surface of the water is horizontal, the parts that we see have a significant inclination leading to a decreased amount of light reflected our way. The water appears darker than the sky, and the horizon is visible. And the steeper the waves, the darker the ocean. As the wave slope increases, say from 0° to 20°, the altitude of the point in the sky from which light is reflected to the eye increases from 0° to 40°. Because the sky is significantly darker at 40° altitude than it is at the horizon (0° altitude), there is even less light incident on the wave's surface, resulting in an even darker sea.

The visibility of the horizon also decreases on overcast days. A cloud cover is far more uniform in brightness near the horizon than the clear sky, leading to a smaller intensity difference between low sky and high sky. As a result there is relatively more light reaching the surface from overhead and the contrast with the low sky just above the horizon is less.

Fig. 3.9A Where is the horizon? Because the water is perfectly flat and our line-of-sight strikes the water at an angle of incidence of about 90°, the water reflects 100% of the skylight beyond it and therefore is the same color and brightness as the sky. The result: sky and sea are indistinguishable and the horizon is invisible. Conditions were clear, not foggy.

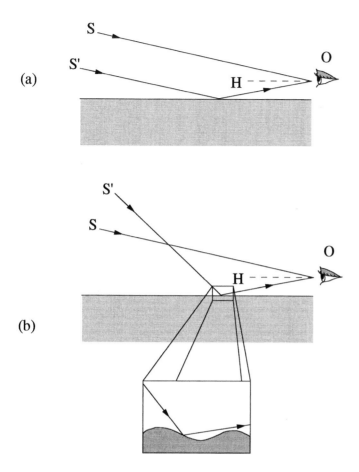

(a)

(b)

Fig. 3.9B How water near the horizon H scatters light from the sky. (a) When the water is flat the water acts like a mirror and the reflected light S′ from some angle below the horizon originates at the same angle above the horizon S. (b) When the water is wavy, reflected light S′ comes from much higher in the sky.

3.10 Wave streaming near the horizon

As we gaze over the ocean, especially through binoculars, we get the impression that the waves are much larger in the horizontal direction (left–right) than they are in the vertical direction (towards and away) (Figure 3.10A). It's as if there are two counter-propagating, co-mingling currents, one from left to right, the other right to left. Yet we see little or no wave motion towards or away from us. Why?

The answer is perspective. Water waves have the same average dimensions in all directions and move randomly. Perspective fore-shortens both angular size and motion in the vertical plane but not in the horizontal (Figure 3.10B). The left–right dimensions and speeds of the waves are preserved while these same properties, towards and away are reduced enormously.

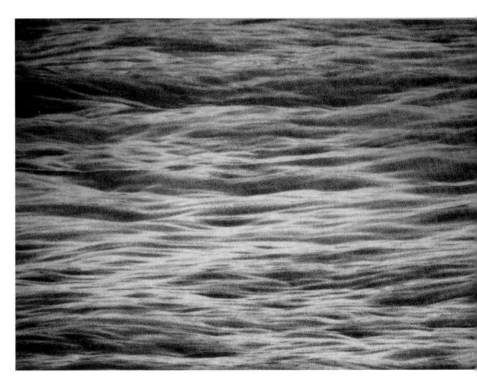

Fig. 3.10A Horizontal structure is characteristic of a nearly horizontal line-of-sight across wavy water. Binoculars show the effect best.

Fig. 3.10B The predominantly horizontal structure of waves and their apparent left–right streaming motion is due to perspective. To the left, isotropic structures like waves (represented here by circles) have the same dimension in both directions. Above, but when viewed at an angle, the waves' left–right size and speed are unaffected by perspective but their towards and away properties are heavily foreshortened.

3.11 Glitter

Looking towards the sun's (or moon's) reflection across a broad stretch of wind-ruffled water we see a brilliant elongated path of sparkling light called glitter (Figures 3.11A, 3.11B). It may extend from the horizon directly below the sun almost to our feet. Seen from the air with a high sun the glitter is elliptical. It is centered on the point on the water that is the same angle below the horizon as the sun's altitude, i.e. at the location where the sun's reflected image would be if the water was flat. As the sun approaches the horizon the glitter becomes narrower although its vertical extent remains the same (Figure 3.11D). When the sun is on the horizon, there is no glitter (Figure 3.11C).

Glitter is the ensemble of countless sun glints. Each glint is an instantaneous flash of sunlight reflected from a wave with just the right slope and position to send the light our way. Because glitter is reflected light, it can be strongly polarized.

The size and elongation of the elliptical glitter depend on the degree of wave development and the altitude of the sun[8-10]. For a high sun the vertical extent of the glitter is four times the maximum wave inclination β (height = 4β) (Figure 3.11E). Why four times? Because reflection doubles the angle (angle of incidence = angle of reflection, deviation $D = i + r = 2i$) and waves face both towards and away from the light source ($\pm$ inclination). From an observer's point of view (Figure 3.11F), this means that we look down at the water and instead of seeing sunlight coming from a single point on the surface as we would if the water was flat, we see sunlight coming from a range of angles that is about four times the average wave inclination. For example, if the maximum tilt is 10°, the vertical range of the glitter will be 40°. The width of the glitter is always narrower than the vertical extent by the sine of the sun's altitude (width = $4\beta \sin a$). This means that for a 10° wave tilt and a solar altitude of 30°, the width of the glitter patch will be 40° $\sin 30° = 20°$ wide.

Fig. 3.11A Glitter patterns on relatively calm water with a low sun.

When the sun's altitude equals twice the maximum wave inclination, the glitter patch reaches up to the horizon and can go no higher. At this point the vertical extent of the glitter is truncated and is no longer equal to 4β but always less (Figure 3.11G). When the sun is lower still and its altitude is less than the maximum wave tilt, the wave shadows itself (Figure 3.11H). Glitter not only decreases in brightness but in vertical extent as well. As the sun sets a point is reached where it no longer shines on the steepest part of the wave, i.e. where the tilt is β. Thus only smaller wave slopes are illuminated and the vertical extent of the glitter patch drops sharply. Finally, when the sun is on the horizon, only the very tops of the highest waves are illuminated and the glitter virtually ceases to exist.

Any number of factors can distort glitter. A change in the wave slope brought about by a gust of wind will cause a sideways bulge in the glitter. An oil slick which damps waves may locally narrow the glitter. The steepness of boat wakes causes bright glitter plumes to reach far outside the normal glitter patch. The same is true in water just over submerged objects or in breakers, both of which produce steeper waves.

When waves travel in a preferred direction, glitter may curve away from the observer[11]. This condition arises because the wave slopes are not uniformly distributed in azimuth and have a systematic tendency to be steeper on one side than on the other. Parallel waves also tend to distort glitter.

Fig. 3.11B Glitter patterns on storm-tossed ocean with steep waves and a high sun.

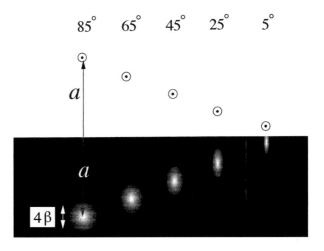

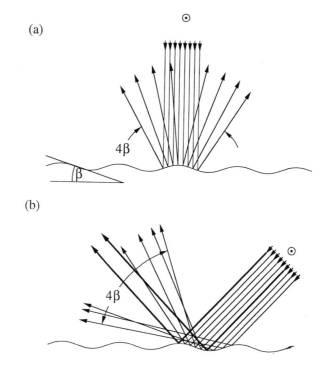

Fig. 3.11D Schematic of glitter as sun sets. When the sun is near the zenith, the glitter pattern is circular and lies directly beneath the observer. As the sun gets closer to the horizon, the vertical height of the glitter is four times the maximum wave slope (4β) but the width decreases.

Fig. 3.11E How waves scatter light from the high sun. As long as the sun illuminates the entire wave, its light will be scattered by four times the maximum wave slope β in the vertical plane.

Fig. 3.11C No glitter is seen when the sun is on the horizon.

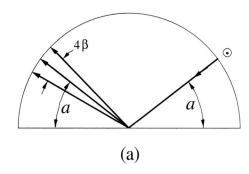

(a)

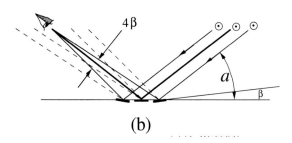

(b)

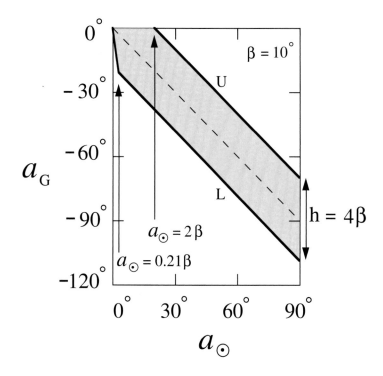

Fig. 3.11F How the observer perceives the scattering of sunlight by water. (a) Waves scatter sunlight through an angle of 4β. (b) The observer also sees light scattered subtending a vertical angle of 4β but the top and bottom rays come from different rays from the sun.

Fig. 3.11G Altitude a_G of the glitter as a function of the sun's altitude $a_\odot$. The altitude of the glitter is negative indicating that it is below the horizon. The upper (U) and lower (L) parts of the glitter are separated by 4β where β is the maximum wave slope. For a maximum wave slope of 10°, the glitter touches the horizon when the sun's altitude is 20° (2β). When the sun is below approximately 2.1° (0.21β) the glitter virtually disappears.

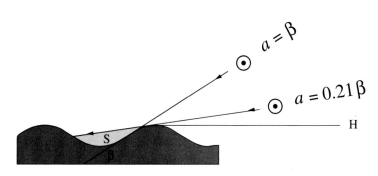

Fig. 3.11H The low sun illuminating only part of the wave surface. The rest of the wave is in shadow (S).

3.12 Skylight reflected from wavy water

To say that the ocean is blue because it reflects skylight is to tell only part of the story (Figure 3.12A). From an observer's point of view, he simply looks out over the water and sees blue light (Figure 3.12B). This light originates from a range of altitudes in the sky. This range of altitudes is 4β. Figure 3.12B is similar to Figure 3.11F; the only difference is that the light paths are reversed and the observer's eye now replaces the sun. Thus the scattering of skylight is in many ways similar to the scattering of sunlight. We can also use Figure 3.12C (a slight modification of Figure 3.11G) to understand quantitatively the way in which skylight is reflected.

Looking down at fairly steep angles, the light comes from a range of points on the sky corresponding to the shape of the glitter patch, i.e. 4β high and $4\beta \sin a$ wide. Towards the horizon (smaller a) a point is reached ($a = 2\beta$) where the horizon limits the amount of light reflected from water waves there. At nearly horizontal lines-of-sight, such as $a = 1°$, the light reflected from the horizon originates many degrees above the horizon. For example, in Figure 3.12C, an observer with a line-of-sight $1°$ below the horizon would see light reflected that came from as much as $12°$ above the horizon. Thus light from the near the horizon sea actually comes from a location that is well above the horizon.

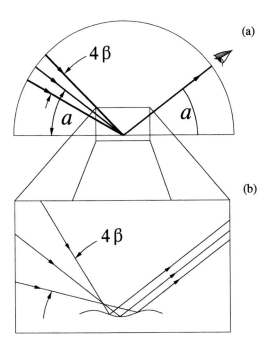

Fig. 3.12B How the observer sees skylight reflected from a wavy surface. Although he may be too distant to resolve individual waves (a) the observer receives light from a vertical range of 4β on the sky. An enlargement of the surface (b) shows how light from different parts of the sky is reflected.

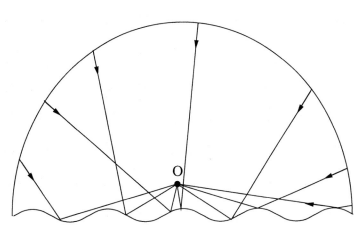

Fig. 3.12A An observer O over water sees skylight reflected from the entire celestial sphere.

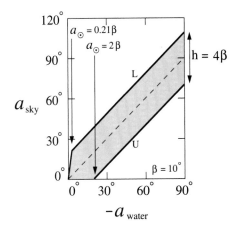

Fig. 3.12C Altitude a_{sky} of the skylight reaching an observer looking down into wavy water at an altitude a_{water}. The altitude of the water is negative, indicating that it is below the horizon. The upper (U) and lower (L) limits of the skylight are separated by 4β where β is the maximum wave slope.

3.13 Moon circles

While the sun's glitter is usually too bright for prolonged enjoyment, the moon's glitter is just right. Glitter is composed of rapidly moving points of light that scurry around the surface describing closed loops (Figure 3.13). These loops are called moon circles[12] and are most prominent when the moon is high and the water gently rippled. A close examination requires a camera but with a little practice we can see them by eye. The loops come and go almost too quickly to follow because their lifetimes are about as long as the eye's response time. Moon circles are common, provided that the source of light is small, bright and nearly overhead. Look for them in swimming pools, hot tubs and even cups of tea.

Time exposures of moon circles reveal their remarkable behavior. At a dark place on the surface, a tiny bright point of light appears, apparently out of nowhere (it is actually a small distorted image of the moon). It pops into view then immediately splits in two[13]. The two points of light glide away from each other and then coalesce, disappearing a fraction of a second after rejoining. Each point may divide in two several times and each pair quickly reunites. Therefore there are always an even number of points.

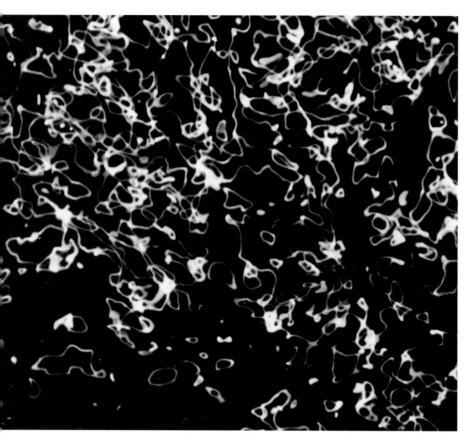

Fig. 3.13 Moon circles seen from a pier. Notice that the glitter patch is composed of myriad closed loops. These come and go almost too fast for the eye to follow.

3.14 Skypools and landpools

Landscape artists and cinematographers love to capture those dancing blue ovals seen in lakes and ponds[14]. These are skypools (Figure 3.14A). They are elongated left–right, and run together creating ever changing ellipses. In very shallow water skypools blend with light from the bottom, creating delicate color and texture transitions (Figure 3.14B).

Skypools are distorted images of the sky and landpools are their complement, distorted images of the shore. The reflected horizon is not a straight line as it would be if seen in flat water but is wavy and distorted. The ensemble is a mixture of both (Figure 3.14C). Towards the distant shore only darkness is seen while near the observer a bright sky dominates.

Skypools are at their best when the observer's line-of-sight is high enough that the entire surface of the wave is seen. In practice this means higher than about 15°. Since this normally occurs only a few meters from the water's surface, one can then completely resolve the skypools and see their form. A photograph will 'freeze' the wave motion.

Light reaching the observer from a wave comes from different parts of the celestial sphere. Light from the far side of the wave comes from near the land (Figure 3.14D) and light from the top and near side of the wave comes from the sky. As the line-of-sight moves down from the top of the skypool where it borders the land to the bottom of the skypool where it again touches the land, the light comes from first higher and higher in the sky, stops (at usually less than 30° altitude in the sky) and then moves downward again. What seems like a smooth continuous image of the sky is actually made of two images from the same part of the sky which fold over and are mirror images of each other[15].

How much sky is reimaged in a skypool? We already know that the vertical extent on the reflected sky is less than about 30°, or twice the maximum wave slope. If the wave slope is less than 15°,

the amount of skypool is correspondingly less. The width of the skypools is determined by the wavelength of the wave and the distance of the observer from it.

Landpools are formed in much the same way as skypools, except that they tend to originate on the slopes of the waves facing away from the observer. Taken together, the optical relation between skypools and landpools is somewhat complementary.

Fig. 3.14A Skypools seen in a deep mountain lake. The skypools appear as horizontally-elongated blue ovals surrounded by light from the opposite shore (in this case rocks and clouds).

Fig. 3.14B Skypools seen in a shallow lake with a mixture of chromatic caustic
network patterns from the bottom and landpools from the shore.

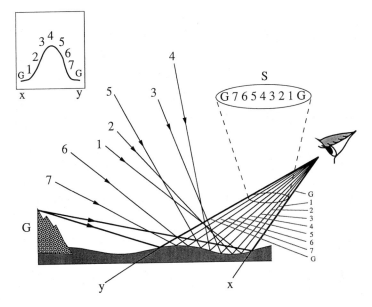

Fig. 3.14D Optics of skypools: skypools (S) are distorted double images of the sky surrounded by distorted images of the shore (G). As the observer looks from the lower edge of the skypool (x) to its upper edge (y), he sees light coming from first higher and higher in the sky (1, 2, 3). Near the center of the skypool the direction reaches a maximum (4) then decreases again (5, 6, 7) until it reaches the shore (G). The inset shows how the angle first goes up then reverses and comes down. Note that the top of the skypool (rays 1–4) gives an inverted image of the sky while the bottom of the skypool (rays 4–7) gives an erect image.

Fig. 3.14C Skypools: as our line-of-sight becomes more horizontal we go from predominantly landpools in the foreground to skypools in the background. Both are most numerous and distinct near their common boundary.

3.15 Shift of reflected skylight towards the horizon

Take a look across a large stretch of rippled water with a single bright cloud well above the horizon. What we see is the reflected light from the cloud on the water at the horizon and not, as we might expect, in the foreground where its mirror image would be (Figure 3.15). Although this reflected light does not form an image of the cloud, its origin is none-the-less clear[11].

To understand this shift we recall that an observer with a low line-of-sight sees skylight on the water near the horizon that has come from many degrees above the horizon. In a clear sky this shift does not draw attention to itself because the entire sky is relatively uniform. But an isolated cloud stands out and we take notice. The shift of reflected cloud light towards the horizon is almost identical in origin to the geometry of glitter. In both cases light from high in the sky is spread out and shifted towards the horizon (Figure 3.9B).

Fig. 3.15 The image of this pinkish sky is displaced toward the horizon. Notice, too, how the cloud bank on the horizon cannot be seen reflected in the water (see also Figure 3.14C).

3.16 The caustic network

An undulating water surface projects patterns of sunlight called caustic networks[16] onto nearby surfaces. Figures 3.16A and 3.16B show the networks in refraction and reflection, respectively. Owing to light absorption and scattering by the water and the peculiar shapes that waves assume naturally, the refractive caustic network is best seen when the water's depth is about five times the waves crest-to-crest distance[17]. A few minutes by the swimming pool will reveal the variety of patterns on the pool bottom formed by this lensing action.

Water's wavy surface can be thought of as a series of positive and negative lenses (Figure 3.16C). The positive lenses focus sunlight onto the bottom creating the bright network. The negative lenses refract light out of the beam and increase the contrast of the network. Since waves propagate across the water, so does the network.

The reflected network is closely related to glitter. Were we to place our eyes at the position of the wall upon which the web falls and look towards the water we would see glitter. Shadow bands and twinkling stars are the atmospheric analogs to caustic networks and glitter, respectively.

Facing the sun and looking onto shallow water, the caustic network on the bottom shows occasional flashes of color, especially when the line-of-sight is nearly horizontal (Figure 3.16D). This color is caused by dispersion at the air–water interface of light from networks on the bottom. When the network is reimaged by surface waves and their light refracted towards the observer, the chromatic aberration of the poorly formed water lens becomes apparent.

Fig. 3.16A The caustic network in refraction.

Fig. 3.16B The caustic network in reflection.

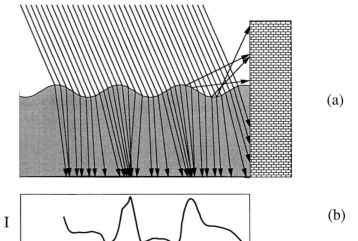

(a)

I

(b)

Fig. 3.16C How the caustic network is created. (a) The water's surface behaves like an array of alternately positive and negative lenses which focus and defocus the light onto the bottom. (b) This creates regions that are in crude focus and are therefore bright. Those out of focus are dark. In reflection, the upper surfaces also acts like an array of curved mirrors which focus or defocus sunlight.

Fig. 3.16D Caustics can show fleeting colors when viewed obliquely.

3.17 Cat's paws

At the first breath of wind over calm water, tiny ruffled patches of dark dimpled water appear, called cat's paws (Figure 3.17). Cat's paws are fleeting areas of capillary waves that spring up almost instantly wherever the wind touches down on the water. Capillary waves are due to surface tension, the springy layer of water at the air–water interface. Cat's paws come and go at the whim of the wind. Like all water waves, cat's paws are generally best seen when the line-of-sight is low and nearly horizontal. When looking more vertically into the water they are harder to see.

Cat's paws darken the water's surface by reflecting less light than flat water. Their novelty lies in the fact that only small, discrete patches of water darken rather than the entire surface. Because surface tension is strong, capillary waves disappear a fraction of a second after the wind stops blowing because their energy source has been removed. In tree-lined lakes or against a mountain seashore, cat's paws may appear lighter than the surrounding water because the distribution of background light is reversed: dark mountains near the horizon and bright sky beyond. Towards the specular reflection of the sun, cat's paws are brighter because they scatter sunlight and add to the glitter.

Fig. 3.17 Cat's paws. Cat's paws are dark regions of the water that show where the wind has touched the surface. The regions are darker than the surrounding flatter water because capillary waves are more steeply inclined to the line-of-sight and therefore the percentage of light reflected is less.

3.18 Slicks and oil-on-troubled-water

Slicks are smooth, shiny areas on gently rippled lakes, on the ocean near shore or in the vicinity of islands (Figure 3.18A). They usually appear lighter in color than the rest of the sea but in the direction of the glitter (Section 3.13), they are darker.

Slicks are usually caused by a thin layer of organic matter produced from fat-bearing organisms[18,19]. Sometimes their origin is in natural or artificial oil seepage. Being more buoyant than water and having a smaller surface tension, the oil floats to the surface and spreads out to form a film only a few molecules thick. Slicks dampen water waves and tend to smooth out the water. As we have already discussed, flat water is brighter than rough water because it reflects a greater percentage of skylight (Figure 3.18B).

The lower surface tension of oil is thought to play some role in wave damping ('oil-on-troubled-water')[20]. The presence of oil takes energy from waves. Since any wave increases the surface area of the water relative to its flat condition, some of the wave's energy is used to stretch the springy oil layer. This energy is not recovered by the wave but instead is converted to heat in the oil, especially for the very short capillary waves. In this manner the film drains the wave energy and leaves a smooth slick patch.

Since slicks are visible only as calm areas in wind-rippled water, they normally cannot be seen on windless days or when the sea is glassy. A breeze of 2–4 knots covers the water with capillary waves and yet does not favor the formation of larger gravity waves. Between 4 and 25 knots, breaking waves disrupt the thin oil layers and the slicks disappear.

Thin films of oil on water produce colorful patterns when viewed by reflected sunlight (Figures 3.18C, 3.18D). The colors result from the interference of light waves reflected at the front and back surfaces of the oil. See also Figure 3.19B.

Impure interference colors are seen on metallic-like scums which Minnaert[11] (p. 212) called ferruginous water. Often noticed in the runoff from mines[21], such scums can curiously be found in pristine circumstances such as a Sierra meadow, Figure 3.18E.

Fig. 3.18A Slicks on the ocean appear as lighter colored areas on the surface. They are caused by oil which suppresses capillary waves (see cat's paws) and therefore flattens the water.

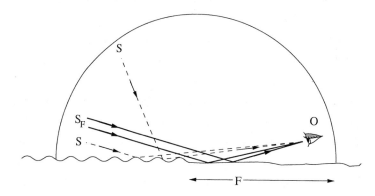

Fig. 3.18B Optics of slicks. The flat water F of a slick is brighter than the surrounding wavy water because it reflects a greater percentage of light, and much of this light comes from the brighter regions just above the horizon.

Fig. 3.18C Oil leaking from the USS *Arizona* at the bottom of Pearl Harbor colors the surface water by interference.

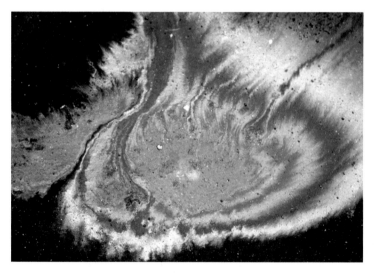

Fig. 3.18D A drop of gasoline on a puddle produced this brilliant display of colors.

Fig. 3.18E Ferruginous water in the apparently absolutely clean third Mono-recess, Sierra Nevada, California.

3.19 Why is foam white?

Niagara Falls is white. So are breaking waves ('whitecaps') and foam from grape soda pop. Regardless of a liquid's true color, its foam is almost always white. Why?

Foaming, bubble-filled water is white primarily for the same reason that clouds are white: scattering of light by spheres. In fact, clouds and foam are opposites of each other – the former being water drops surrounded by the air, the latter are air 'droplets' surrounded by water. They are remarkably similar in the way they scatter light (Figure 3.19A). Foam contains a wide range of spherical bubbles, ranging from less than the wavelength of light up to several millimeters. Although each bubble scatters light with a well-defined and even colorful manner, when taken together with other bubbles the result is an achromatic sum, i.e. white. The same thing is true when colored material is ground into powder: regardless of its intrinsic color, many finely divided substances appear white.

Seen close up, foam is not always pure white because other optical effects may be present. A close look at sea foam that lingers on the shore reveals a wide spectrum of colors, all due to thin film interference (Figure 3.19B).

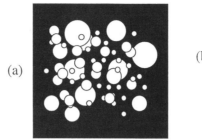
(a)

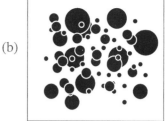
(b)

Fig. 3.19A Structure of foam: foam (a) is just the opposite of a cloud. It is made of air bubbles surrounded by water while a cloud (b) is made of water droplets surrounded by air.

Fig. 3.19B Interference colors in foam. The colors are due to thin film interference and the color they show depends on their thickness and the angle that the observer's line-of-sight makes with the surface.

3.20 The wet spot

A splash of water on a shirt looks dark (Figure 3.20A). Moist ground appears dark. Wet hair looks dark. In fact nearly all wet surfaces are darker than when dry[22]. Why?

There seem to be two explanations for the darkness of a wet spot, depending on whether water penetrates the underlying material or not. The simplest case is when a thin layer of water overlays a non-porous substance, for example a piece of concrete (Figure 3.20B). Light passes through the water and strikes the concrete below where some of it is absorbed. The rest is scattered diffusely in all directions. Some of the scattered light strikes the air–water interface from below at large enough angles of incidence to be totally reflected back downward. This internally reflected light again strikes the concrete where more of it is absorbed. In this manner, light entering through the film of water is repeatedly scattered and absorbed[23]. When the light finally leaves the wet surface it is dimmer than light scattered by the surrounding dry surfaces and therefore appears dark by contrast.

On porous substances like sand or fabric, a different mechanism is at work (Figure 3.20C). Viewed under a microscope, almost everything shows a great deal of surface structure, much of which is close to the size of the wavelength of light; sand is gritty and fabric fuzzy. Such features scatter and diffract light efficiently regardless of the material's intrinsic color. For these reasons there are always two components of light coming from every dry surface: reflected light characteristic of the material's color and scattered light from its surface texture. It is this additional scattered light that lightens the surface and dilutes the object's true color.

When the surface is moistened, the scattering structures become coated with a thin layer of water. Light striking the surface now passes from air ($n=1.000$) to water ($n=1.333$) and finally into the surface below (where n is much greater than 1.3333, say 2.000). The difference in index of refraction between water and the substance is also less than that between air and the substance. The amount of light reflected at any interface depends on the ratio of the two indices. By changing this ratio from 1.000/2.000 to 1.000/1.333, an antireflection condition much like the one used on camera lenses is created. This coating reduces the amount of light scattered from the surface, thereby making it appear darker and closer to its true (reflected) color.

Light coming from such a spot is darker than normal, so the light must be going somewhere. It is scattered deeper into the material more efficiently than when the material is dry[24]. With deeper penetration, the light naturally strikes many more surfaces before escaping. With each particle or bit of matter struck by the light, a certain fraction of the light is absorbed. By the time any light finally gets out, it has been significantly attenuated.

Fig. 3.20A Wet spots. They are always darker than the dry surface.

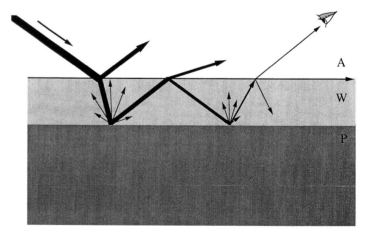

Fig. 3.20B Wet spot optics: optics of wet spots on nonporous substances. A: Air, W: Water.

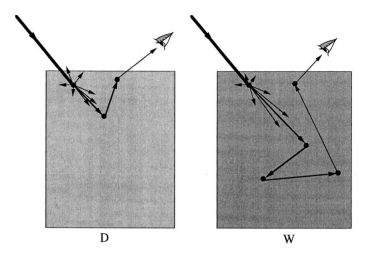

Fig. 3.20C Wet spot optics: optics of wet spots on porous substances.

3.21 Shadows on water

Look at your shadow on quiet water. Can you see it? If so it means that the water contains suspended particles. Perhaps you can see the three-dimensional character of your shadow converging to the antisolar point. In the very clearest of deep water, no visible shadow is produced.

Particles in water scatter light and cause it to be brighter in the sunlight and darker in the shadow. Contrast makes the shadow visible. If the water is fairly clear, you can see your three-dimensional shadow reaching down into the water. It will assume the triangular shape analogous to mountain shadows and the Spectre of the Brocken. If the water is totally opaque, you see only a simple shadow like the one you normally cast on the ground.

When looking into the shadow of another object, it is easier to see subsurface structures because light coming from the shaded region has a much weaker component of reflected surface light. Thus, image contrast improves in the absence of reflected surface light. That's why it is easier to see fish in the shadows of overhanging trees or on the shady side of a boat.

3.22 Aureole effect

Standing within a few meters of fairly smooth water upon which your shadow is cast, look at the water around the shadow of your head. You may see dancing rays radiating from your shadow (Figure 3.22A). This sparkling crown is called the aureole effect[25]. Rays from the aureole can be seen many degrees from the anti-solar point if the observer's viewpoint is elevated. The aureole effect can occasionally be observed from aircraft.

The light and dark rays of the aureole are caused by water waves that focus and defocus sunlight down through the water (Figure 3.22B). Though focusing always occurs, it is not evident unless there are some suspended particles present to scatter the light. Where converging light increases the average brightness, we see bright rays. Defocused (diverging) light gives dark rays.

The aureole effect's radial structure is due to perspective. The focused ray bundles are more-or-less parallel, But because our line-of-sight is parallel to, and lies within the ray bundles, they appear to converge to the antisolar point. The antisolar point is within the shadow of our head and our shadow appears surrounded by twinkling, radiating rays.

Fig. 3.22A Looking into water toward the antisolar point with the sun to our back reveals the aureole effect of self-centered shadows.

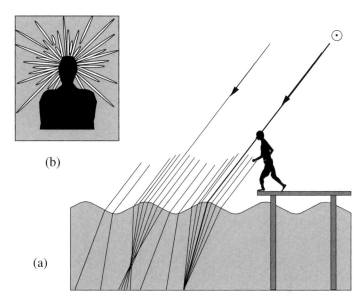

Fig. 3.22B The aureole effect arises from the caustic network being projected into water that is turbid but clear enough that we see below the surface. Since each bundle of rays going down into the depths, though converging on themselves, is parallel to the others, they appear to converge to the antisolar point as a result of perspective.

3.23 Other reflections on water

Long thin structures like sailboat masts seen reflected in gently-ruffled water often distort into sinuous curves[12,26,27]. They break into closed nested loops containing structure that is reminiscent of skypools (Figure 3.23A). As we watch the images they weave and change, all the while keeping time to the waves. The circular patterns of masts are similar to moon circles except in one sense. While moon circles are formed from a point-like image and trace out their closed paths with time, the circular patterns of masts originate from long thin sources.

When the line-of-sight is fairly low, a series of parallel wave trains (like from a boat wake) causes the landscape to be repeatedly imaged (Figure 3.23B). Each image comes from nearly the same part of each wave, usually near the crest. Like skypools, the images are folded.

The reflected images of buildings and bridges in wind-ruffled water reproduce the vertical lines in the original scene while distorting the horizontal lines to the point of destroying them altogether (Figure 3.23C)[28]. This effect is analogous to the sun's image smeared vertically into glitter. A smeared vertical line still looks like a line. But a horizontal line smeared vertically is destroyed.

The clarity with which an image is reflected from water depends on the water's roughness; images may be distorted in gently-rippled water but may still be recognizable. In rough water no identifiable image can be seen. This is evident in Figure 3.23D in a before-and-after sequence. Just before a gust of wind came along and roughened the surface, the images of the masts were obvious. After the cat's paws and short wavelength waves developed, there was no sign of the masts.

Fig. 3.23A Reflections of sailboat masts.

Fig. 3.23B Multiple distortions arise in the presence of wave trains. (Photo by G. Ladd)

Fig. 3.23C Anisotropic reflection of a bridge on a river where vertical features are distinct while horizontal ones are diffuse.

(a)

(b)

Fig. 3.23D (a) Distorted but still recognizable images of sailboats and their masts in gently rippled water. (b) A gust of wind has covered the surface with capillary waves and destroyed the reflected images.

References

1 Bohren, C., 1982, 'Colors of the sea', *Weatherwise*, **35**, 256–60.

2 Bohren, C., 1983, 'More about colors of the sea', *Weatherwise*, **36**, 311–16.

3 Smith, R.C. and Tyler, J.E., 1967, 'Optical properties of clear natural water', *Journal of the Optical Society of America*, **57**, 589–601.

4 Tyler, J. E., 1965, 'In-situ spectroscopy in ocean and lake waters', *Journal of the Optical Society of America*, **55**, 800–05

5 Smith, R.C. and Baker, K. S., 1981, 'Optical properties of the clearest natural waters', *Applied Optics*, **20**, 177–84.

6 Plass, G.N., Humphreys, T.J. and Kattawar, G.W., 1978, 'Color of the ocean', *Applied Optics*, **17**, 1432-1446.

7 Preisendorfer, R.W., 1976, *Hydrologic Optics, I*, Sec 3.8, 34. US Department of Commerce.

8 Hulburt, E.O., 1934, 'The polarization of light at sea', *Journal of the Optical Society of America*, **24**, 35–42.

9 Cox, C. and Munk, W., 1954, 'Measurement of the roughness of the sea surface from Photographs of the Sun's Glitter', *Journal of the Optical Society of America*, **44**, 838–50.

10 Goodell, J.B., 1971, 'The appearance of the sea reflected sky', *Applied Optics*, **10**, 223–25.

11 Minnaert, M.M., 1954, *The Nature of Colour and Light in the Open Air*, Dover, New York.

12 Lynch, D.K., 1985, 'Reflections on closed loops', *Nature*, **316**, 216–17.

13 Longuet-Higgins, M.S., 1960, 'Reflection and refraction at a random moving surface. I. Patterns and paths of specular points', *Journal of the Optical Society of America*, **50**, 838–44.

14 Farber, D., 1991, *Reflections on a Trail Taken*, Godin Publishing, Boston.

15 Thomas, D.E., 1980, 'Mirror images', *Scientific American*, **243**, 206–28.

16 Berry, M., 1981, 'Singularities in waves and rays', in *Physics of Defects*, Les Houches Lectures XXXIV, R. D. Balain, M. Kleman and J-P Poirier, Eds., North Holland, Amsterdam, 453–543.

17 Schenck, H., 1957, 'On the focussing of sunlight by ocean waves', *Journal of the Optical Society of America*, **47**, 653–7.

18 Dietz, R.S. and LaFond, E.C., 1950, 'Natural slicks in the ocean', *Journal of Materials Research*, **9**.

19 Ewing, G., 1950, 'Slicks, surface films, and internal waves', *Journal of Materials Research*, **9**, 161.

20 Van Dorn, W.G., 1974, *Oceanography and Seamanship*, Dodd Mead, New York.

21 Kratochvil, D. and Volesky, B., 1998, 'Biosorption of Cu from ferruginous water by algal biomass', *Water Research*, **32**, 2760 (and references therein).

22 Angstrom, A., 1925, 'The albedo of various surfaces of ground', *Geogr. Ann.*, **7**, 323.

23 Lekner, J. and Dorf, M.C., 1988, 'Why some things are darker when wet', *Applied Optics*, **27**, 1278–80.

24 Twomey, S.A., Bohren, C.F. and Mergenthaler, J.L., 1986, 'Reflectance and albedo differences between wet and dry surfaces', *Applied Optics*, **25**, no. 3, 431–7.

25 Jacobs, S., 1953, 'Self-Centered Shadow', *American Journal of Physics*, **21**, 234

26 Gold, T., 1985, 'Riddle of reflections in the water', *Nature*, **314**, 12.

27 Berry, M.V., 1987, 'Disruption of images: the caustic touching theorem', *Journal of the Optical Society of America*, A, **4**, 561.

28 Tricker, R.A.R., 1965, *Bores, Breakers, Waves and Wakes*, American Elsevier, New York.

29 Invert the picture.

4 Water drops

It was late afternoon as we approached Tok Junction on the Alaskan highway. Someone yelled: 'Look at that strange rainbow'. Strange indeed, it had three, maybe four arcs which intersected at unfamiliar angles! What we were witnessing was a rare reflection phenomenon – the superposition of a normal double rainbow with a fainter double bow formed from rays first reflected off a lake. (See Section 4.9.)

Adapted from the authors' observing books

RAINBOWS AND THEIR KIN

4.1 Sparkling dewdrops

On early sunny mornings a dew-covered lawn displays eye-catching sheens. The grass scintillates with brilliant sparkles of light. Each sparkle originates at a water drop. Some drops show intense white glints, others gleam with the colors of blue, yellow, orange, or red (Figure 4.1A).

Move close to a particular drop, so close that you can no longer focus on it. Relax your eyes. Avoid your own shadow so that the drop remains illuminated and examine the out-of-focus glint (Section 7.17). Some viewing angle will be found where the sparkle takes on color. Shift a little sideways and the color will change. Now notice the following: redward there is an abrupt fall off of intensity while blueward the light fades slowly without any distinct cutoff. What you are experiencing is the rainbow from a single drop of water.

With careful positioning a rainbow display can be discovered in a single drop of tree sap, Figure 4.1B.

4.2 Observations of rainbows

Rainbows are those colored arcs seen on falling water drops. Brightest is the primary bow which is centered on the antisolar point. Its radius is about 42°. There is a fainter secondary arc at 51°. When both the primary and secondary are present we have a 'double rainbow' (Figure 4.2A). Viewed from an elevated place (airplane) a rainbow can inscribe a complete circle.

Fig. 4.1A A neighborhood lawn sparkles with light from dewdrops. Inset on left contains a rainbow fragment from a single drop produced by out-of-focus viewing (Section 7.17). Inset on right reveals single drop landscape images (Section 4.16).

Fig. 4.1B A droplet of tree sap, or balsam, produced this out-of-focus rainbow fragment. (Photo by G. Ladd)

How frequently we see rainbows depends on whether the sun is low or high in the sky, a factor which depends on the time-of-day, the latitude, and season of the year. For an observer on the ground, the antisolar point must be higher than −42° elevation in order that a primary bow be above the horizon. Consequently rainbows are more likely in the early morning or late afternoon. Rainbows also favor the thunder showers of summer, with their broken sunlit skys, as opposed to the complete overcast condition that accompanies winter storms.

The intensity of rainbow colors vary for several reasons[2]. One condition is the strength and hue of the ambient sunlight which creates it. Near sunset, when the blue-greens have been removed by atmospheric scattering, a rainbow is decidedly red (Figure 4.2B). Another factor is the size of the water drops (Figure 4.2C). Large drops produce the most vivid colors. As size diminishes, wave interference broadens the dispersed light to cause an overlap of the colors. Obviously a dark background is favorable for rainbows while the presence of bright clouds is unfavorable.

Sunlight is about one million times brighter than moonlight. A rainbow produced by the moon invokes scotopic vision which is devoid of color sensation. A lunar bow is therefore usually colorless. Similarly we may be color blind to rainbows produced by artificial lamps at night[3].

Rainbows can be found in unexpected places. Fragments of rainbows are found in meadows on dew-covered spider webs (Figure 4.2D). These spectra reveal that cobwebs litter the ground in surprising numbers. At the beach we see 'surf bows'; from thoughtless drivers we dodge 'road spray bows'; and there is the 'marine bow' at the prow of a ship. Most reliable is the 'garden hose bow' which helps to inspire us to yard work. This is the one rainbow accessible to us on any sunny day. Finally, we may vacation in Yellowstone hoping to see a 'geyser bow', or in Yosemite where 'mist bows' frequent its cascades. The sharp-eyed reader will undoubtedly enlarge this list.

Rainbows are not confined to visible wavelengths; they exist in both the near infrared[4] and ultraviolet (Figure 4.2E). Incidently, someone who has had a cataract operation, and thereby possesses aphakic eyes, will see the broad ultraviolet component to a rainbow that others miss (Figure 4.2E). His or her perception is thereby expanded over that of normal eyesight.

Infrared and ultraviolet bows are weak, but for differing reasons. An infrared bow is attenuated by the increasing opacity of water at wavelengths greater than 700 nanometers. On the other hand, while water drops are almost perfectly transparent down to the terrestrial transmission cutoff at 300 nanometers, the fraction of ambient light received directly from the solar disk diminishes at short wavelengths. Atmospheric scattering redistributes sunlight to brighten the sky at the expense of the direct component.

Not located at a particular place, a rainbow exists only as a direction. Whether the water drops are one meter or a kilometer away, the primary rainbow always appears at an angle of 42° around the antisolar point. As a consequence, side-by-side

Fig. 4.2A Double rainbow with primary and supernumerary graces the sky over Flagstaff, Arizona. Notice that the supernumerary fades toward the ground (Section 4.5). Alexander's dark band is the darkening of the sky between the primary and secondary, so named for the Greek sage who first chronicled it[1].

observers each see a different bow. Parallax can play tricks with rainbows[5]. Carried to the extreme, each of our eyes sees a different bow. In the case of the 'swimmer's bow', which is formed less than a meter away in the swimmer's spray, a double primary is experienced, one for each eye[6].

Any shadow rays from clouds that happen to intercept a rainbow must cross perpendicular to the bow. Such anti-crepuscular rays (Section 1.9) will terminate at the arc to create the illusion of a 'rainbow wheel' (Figure 4.2F). Their radial convergence toward the antisolar point is a consequence of perspective. John Constable's 1831 painting *Hampstead Heath in a Storm* contains a nice rainbow wheel.

Fig. 4.2C The large drops of a thunder shower yield the best colors.

Fig. 4.2B Red rainbow near sunset. The blue and green light has been attenuated by the atmosphere. (Photo by O.R. Norton)

Fig. 4.2E Garden hose bow photographed through narrow band filters at 320 nanometers in the ultraviolet and 650 nanometers in the red.

Fig. 4.2D Fragment of a primary rainbow seen in the dew of an orbweb.

Fig. 4.2F Rainbow wheel as seen over the Grand Canyon. (Photo by G. Ladd)

4.3 The primary rainbow

This arc of 42° radius has a width of about 2°. Color runs blue to red outward from the bow's center. How often the inattentive artist reverses these colors.

A ray of light that passes through a water drop with one internal reflection is refracted and dispersed, somewhat like in a prism, to produce a spectrum. There are many routes that such a ray can take through a drop, yet the rainbow in the sky appears at the particular angle of 42° and the colored light is concentrated there. Why?

Assume for the moment the sun is a point source and consider the paths of a bundle of parallel light rays falling on a water drop. In Figure 4.3A each incident ray has been spaced so that it represents an annulus containing equal energy. Angle D is the deviation between the incoming and outgoing rays that experience one reflection inside the drop. For a central ray, which reflects back on itself, the light is deviated by 180°. As the center line distance X goes away from zero (central), deviation D at first decreases rapidly, attains a minimum value of approximately 138°, after which D increases once again for the outermost rays as X approaches unity (Figure 4.3B). At the minimum deviation angle, incoming rays from a range of X are compressed into a narrow bundle about the 138° direction. This enhancement creates the primary bow. Reckoned from the antisolar point the rainbow has a radius of 180° – 138° = 42°. With the sky filled with drops, each one contributes a sparkle as it falls through the region of space near 42° from the antisolar point and the light of these sum to produce the primary arc.

Each color has a different minimum deviation angle. For red light it is 42.25°, for blue 40.91°. Table 4.3 lists rainbow angles and their dependence on wavelength.

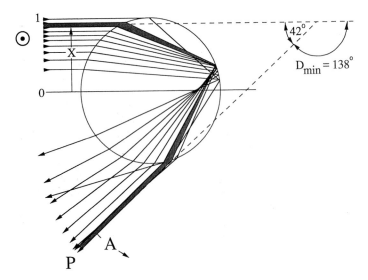

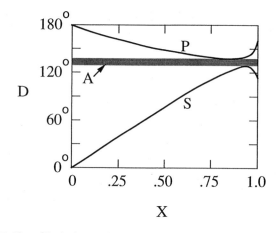

Fig. 4.3B Plot of deviation angle D versus ray distance X from the sun–drop line. P is primary, S secondary, and A Alexander's dark band.

Fig. 4.3A Ray diagram for a single wavelength showing how exiting beams from a water drop pile up about the minimum deviation angle to create the brightness of a primary bow. Since a drop is spherical, an identical diagram may be drawn for rays outside the plane of this paper by rotating into this chosen plane. The circle with a dot in it represents light from the sun; P is the primary rainbow; A is Alexander's dark band; D_{min} is the minimum deviation angle; X is the distance of any ray from the sun–drop centerline. Rainbow angle is 180° – D_{min} = 42°.

Table 4.3. *Rainbow angles for the minimum deviation rays as they depend on wavelength and the index of refraction of water n.*

Wavelength (nanometers)	Index n	Primary angle (degrees)	Secondary angle (degrees)
1000	1.3277	42.86	49.49
900	1.3285	42.73	49.70
800	1.3294	42.60	49.92
700	1.3309	42.38	50.34
650	1.3318	42.25	50.58
600	1.3335	42.01	51.02
550	1.3344	41.64	51.68
500	1.3364	41.27	52.33
450	1.3411	40.91	52.99
400	1.3440	40.51	53.73
350	1.3501	39.66	55.26
300	1.3532	38.72	56.80

Other transparent media such as ocean saltwater, the balsam of trees, and various oils all have their own indices of refraction and consequently different values for the rainbow angle. An alert photographer once recorded the juxtaposition of freshwater and saltwater rainbows. He called this esoteric happening a 'broken bow', something you and I may never encounter[7] (Figure 4.3C).

How pure are the colors of a rainbow? By purity we mean to what degree do the colors overlap? Each color, or wavelength λ, has a minimum deviation angle given in Table 4.3. For blue ($\lambda = 400$ nanometers) the angle is 40.51°. The sun has a diameter of 0.5°. Each spectral element is spread out by this amount. In the blue 0.5° corresponds to about 40 nanometers, so 400/40 = 10, and we say the solar disk limits rainbow spectral purity to 1 part in 10. In the red it is 1 part in 5. Because the solar disk is circular, its effective width is somewhat less than half a degree and spectral purity should be a bit better than our estimate. But there remains another factor which smears the rainbow spectrum: all rays do not emerge at the minimum deviation angle. Taking both the solar disk and minimum deviation into account the spectral purity is less than one part in four in the blue (Figure 4.3D).

Fig. 4.3C A broken rainbow occurs when a surf bow meets up with an atmospheric rainbow. The rainbow angle for seawater is 0.8° smaller than for freshwater. (Photographed at sea by J. Dijkema)

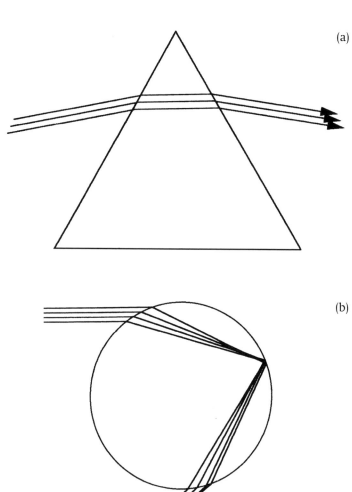

(a)

(b)

Fig. 4.3D A bundle of monochromatic, parallel rays entering a prism emerge parallel (a), while a similar bundle diverges from a water drop (b). This divergence, together with the 0.5° width of the solar disk, explains why rainbow colors are impure.

4.4 The secondary rainbow

Often a second and fainter rainbow is seen outside the primary at about 51°. Its colors are reversed from the primary, being red to blue outward from the antisolar point, and the bow is almost twice as wide. Because it is both fainter and broader, the secondary may not be evident.

A secondary bow results from two internal reflections in the water drops (Figure 4.4). Because some light is lost at each reflection, the secondary's brightness is about 43% of the primary's. Table 4.3 also gives the minimum deviation angles for the secondary.

4.5 Supernumerary bows

Once in a while a faint narrow bow is tucked in close under the primary bow. In rare instances a whole series of such bows can be found there, especially along the horizontal parts of the arc[8]. These are supernumerary rainbows.

They are especially obvious in the garden hose bow when the nozzle is set for a fine mist. Sometimes a supernumerary can even be found attached to the outside of the secondary (Figure 4.5A). But strangely most natural rainbows display no supernumeraries at all.

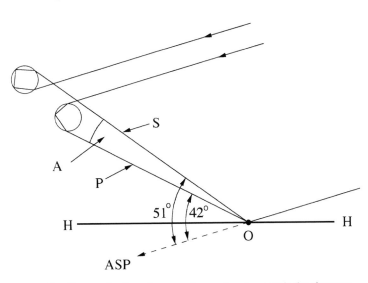

Fig. 4.4 Sky diagram for the primary and secondary bows. O is the observer; H–H the horizon; P the primary bow; S the secondary; A Alexander's dark band.

Fig. 4.5A Garden hose bow displaying a rare supernumerary (arrow) outside the secondary bow.

Unlike the primary and secondary rainbows, which can be explained by geometrical optics, we need the wave theory of light to understand supernumeraries. They are caused by interference between two parts of a light wave that have passed through a drop along slightly different paths (Figure 4.5B). The two rays emerge at nearly the same position and interfere. Several orders of interference may be visible.

Why are supernumeraries so elusive in a natural rainbow yet readily seen in the garden hose bow? The answer lies in the fact that supernumerary spacing depends critically on water drop size. Figure 4.5C shows how light intensity is predicted to vary across the rainbow angles for spherical drops[9]. As drop dimensions increase the supernumerary spacing closes up. In a typical rain shower there will be a range of drop sizes and these smear together the interference patterns to render them invisible. A hose nozzle, on the other hand, lets through only drops having selected radii and the supernumeraries are nicely revealed.

When supernumeraries are present in a rainbow, why are they mainly confined to the crest of the arc? Alistair Fraser[8] has introduced another variable into rainbow theory: the flattening of large rain drops as they fall through the air. At the risk of being hit in the eye, gaze upward during a thunder shower and you will see that the biggest drops are misshapen and oscillate wildly. Obviously most of these cannot contribute to a rainbow. High speed photography reveals that even the smaller drops are flattened somewhat[10-13]. On average a vertical cross-section through these smaller sized drops will be elliptical, while in a horizontal cut they are circular. Of course tiny drops are circular about both axes. Taking into account flattening, a diagram of rainbow light intensity shows that between 0.2 and 0.3 millimeters radii the ripple spacing hardly changes, and for this range of drops their interference patterns add to become visible. At the foot of a rainbow, all horizontal slices through a drop are circular and there is no supernumerary.

One result of this drop flattening theory is that the spacing between all supernumeraries should be constant and about 0.7°. A second feature is that, for appreciable flattening, the radius of the primary arc ought to shrink by several degrees. This means the shape of a thunder shower bow would be distinctly non-circular. This is not observed, presumably because such flattened drops are oscillating and fail to direct light into the rainbow.

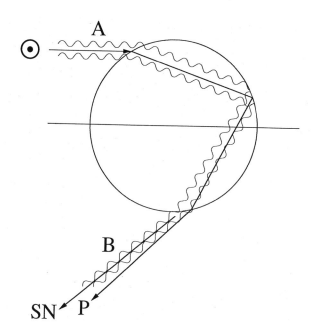

Fig. 4.5B Source of the supernumeraries SN. Two solar (circle with dot) rays incident on a drop will emerge with their relative phase changed because of differing path lengths through the water. If the rays are initially in phase as at A, there is an angle such that they will exit out of phase and thus cancel as at B. Increasing deviation will cause alternating constructive and destructive interference which is the observed pattern of supernumeraries.

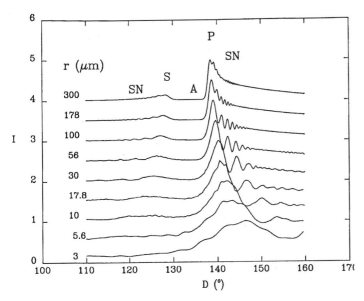

Fig. 4.5C Theoretical scattering diagram for rainbows; spherical particles assumed. I is the intensity on a logarithmic scale; r is drop radius in micrometers (μm); P = primary; S = secondary; SN = supernumeraries; A = Alexander's dark band. D is the deviation angle (recall rainbow angle = 180° – D°).

WHAT SHAPE ARE RAINDROPS?

Anyone who has studied physics knows that the mythical waterdrop with a pointy tail can never exist in nature – even momentarily. Surface tension will prevent a sharp cusp from forming on any water surface. Instead, the powerful forces of surface tension try to minimize surface area which normally leads to spherical drops.

But if the drops are fairly large, as happens in a thunder shower for example, those over a couple of millimeters in size become decidedly flattened as they fall through the air. To see this simply look directly up during such a shower. (Yes, you will get wet, but bravery is called for here.) As the drops rush toward your eyes they appear distorted and oscillate wildly. We have noticed that artificial fluorescent lightning emphasizes the effect by its strobing action.

4.6 Tertiary bows

Rainbows arising from three, four, and more internal reflections are termed tertiary, quaternary, etc. bows and have been observed in the laboratory using laser illumination on single droplets. A tabulation of rainbow angles and predicted intensities up to order 20 has been published[14]. Table 4.6 gives the predictions of rainbow angles and intensities for the first five orders.

The third internal reflection should cause a bow in the sky within (180° – 137° =) 43° of the sun with an intensity about half that of the secondary. A fourth order bow is found at about the same place with even less intensity. In principle the blue component of the fifth order bow might be visible as it falls within Alexander's dark band (see below)[15]. Some years ago Jearl Walker wrote about the impossibility of seeing a high order bow in nature[16]. As a result several sightings came to his attention, one or two possibly legitimate. He concluded that under exceptional circumstances the third and fourth order bows might be observable. But in general these higher order bows are so faint, broad, and so unfavorably positioned relative to the glare of the sun, or are superimposed on the brighter primary or secondary bows, that we believe them to be invisible in the landscape.

Table 4.6. *Rainbow angles for the first five orders in red light. Calculated intensity is relative to the primary[13].*

Order	Width (degrees)	Angle (degrees)	Intensity
1	1.72	42.38	1.00
2	3.11	50.37	0.43
3	4.37	137.52	0.24
4	5.58	137.24	0.15
5	6.78	52.92	0.10

4.7 Alexander's dark band

The area between the primary and secondary bows is noticeably darker than the surrounding sky. Called 'Alexander's dark band', it is named for the Greek sage Alexander of Aphrodisias who in AD 200 described it in his chronicles[17]. This geometric band is an area devoid of rays which have passed through water drops (Figure 4.7). Light reflected from the surface of drops can reach this band, but those rays are relatively faint. In the case of a dewdrop spectrum, Alexander's zone will appear very dark indeed because there are no competing sources of scattered light at this angle except for the surface reflection.

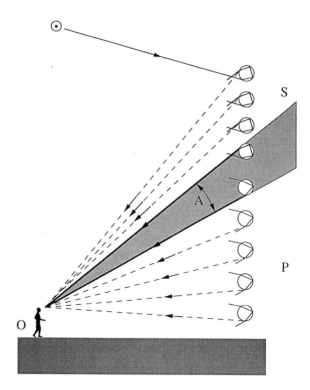

Fig. 4.7 Why the sky is relatively dark between the primary and secondary arcs (Alexander's dark band). Permitted rays (rays greater than those corresponding to the minimum deviation angle) are indicated with rainbow angle rays drawn connecting to the eye. Imagine an array of such drops from a rain sheet and you will find that no rays between 42° and 51° reach the eye. O is the observer; P are drops producing the primary bow; S drops of the secondary; A is Alexander's dark band.

4.8 Polarization of rainbows

View a rainbow through a polaroid and notice how it may be extinguished (Figures 4.8A, 4.8B). But not everywhere simultaneously; the plane of polarization is always tangent to the arc.

Measurements with a polarimeter show that the rainbow is polarized up to 94%. This polarization arises at the internal reflection in the water drop which is near the Brewster angle. The bright sky inside the primary bow is also tangentially polarized since most of this additional scattered light has taken a similar path through the droplets.

Fig. 4.8A Rainbow light and the adjacent sky may be extinguished with a polaroid.

Fig. 4.8B Rotate the polaroid 90° and the light of the rainbow is passed by the polaroid.

4.9 Reflection rainbows

It comes as no surprise that rainbows can be seen reflected in
water. More interesting are rainbows from light which has first
been reflected off a lake before reaching the drops (Figure 4.9). In
this case four arcs are possible: the direct double rainbow and a
reflected, inverted, double bow. See plate 1–9 of Greenler[4] for a
nice photograph of this spectacle.

   The reflectivity of water approaches 100% near grazing incidence (Figure 3.2B). This means that under conditions of a low sun
and calm water, a rain sheet is exposed to two suns of almost equal
brightness. As a consequence two sets of rainbows are created
about the two antisolar points.

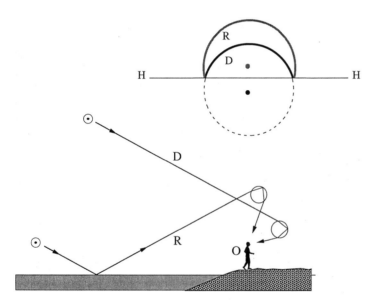

Fig. 4.9 Circumstances of the reflection-light rainbow. To the observer there
are two antisolar points, one below and one above the horizon as shown. The
direct double rainbow and the reflected-light double form in the usual
manner about these two antisolar points. D is the direct and R the reflected
component of sunlight; O the observer; H–H the horizon.

4.10 Fogbows

A pale white rainbow is sometimes seen in clouds and fog banks (Figures 4.10A, 4.10B). This fogbow is low in contrast and has a width greater than the 2° of a normal rainbow[9,18].

As mentioned earlier, large drops produce the most vivid color in rainbows. As the drops become smaller, wave interference increases to cause an overlap between rays of differing wavelengths (see Figure 4.5C). At a drizzle drop diameter of 0.1 millimeters color is not perceptible save for a reddish tinge at the outer rim. For still smaller droplets (diameters less than 0.050 milli-

meters) all hint of color vanishes and we have the uncommon fogbow. In most clouds the drop sizes are too small to create even a fogbow.

Fig. 4.10A This fogbow arises from drops less than 0.05 millimeters in diameter, or about a tenth the size of drops that produce colored bows. Seen here on Mauna Kea at about 4200 meters.

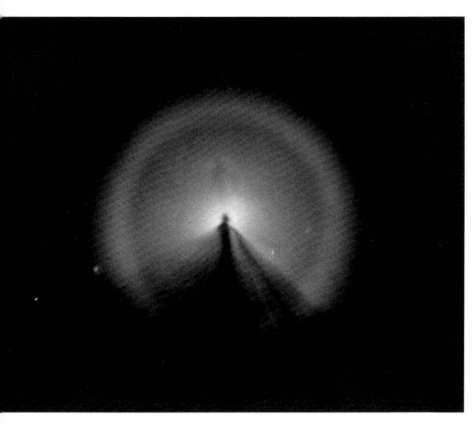

Fig. 4.10B Fogbow with spectre shadow and a glory. Above the observer's shadow is a faint, ghostly replica. This is a displaced version arising from reflected light off wet pavement to the rear, exactly analogous to the reflected rainbow diagramed in Figure. 4.9.

4.11 Cloud contrast bows

While flying over a deck of clouds, one becomes aware that within an angle of about 42° of the antisolar point cloud structure becomes curiously accentuated. The effect is subtle and requires the motion of the airplane to enhance its visibility. This structure is called a 'cloud contrast bow'[19] (Figure 4.11).

Drop sizes in cumulus and stratus clouds range from 0.001 to 0.050 millimeters, implying an absence of rainbows[20]. Some fraction of the light from cloud tops has been scattered only once. This light has entered drops, then left them and the cloud without further interaction. This means there will be a brightening at the rainbow angle, but as with the fogbow, the enhancement will be broad and colorless. On the other hand, light from shadows among the folds in a cloud, which provides the contrast to see cloud forms, is more likely to be multiply-scattered and lacks such an enhancement at any viewing angle. For this reason the contrast between cloud and shadow reaches a maximum at the rainbow angle to produce the cloud contrast bow.

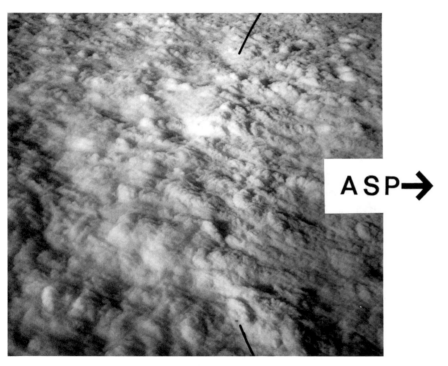

Fig. 4.11 The cloud contrast bow is seen by the airplane passenger as augmented detail in the structure of stratocumulus cloud. Found interior to the rainbow angle, shown, motion clarifies the effect over a still photograph. ASP = antisolar point.

FORWARD AND BACKSCATTER PHENOMENA

4.12 Heiligenschein

On a sunny morning before the dew has evaporated, look for a bright colorless glow a few degrees across around the shadow of your head (Figures 4.12A, 4.12B). This is the heiligenschein, or 'holy light'.

The heiligenschein is a retro-reflection that originates when dewdrops focus sunlight on blades of grass. Sunlight is brought to a crude focus behind a water drop at a distance of about 0.7 times the drop radius (Figure 4.12C). If the drop does not lie close to a surface, the light converges and then continues on its way, that being the end of it. But if the concentrated light leaving the drop immediately strikes a blade of grass (Figure 4.12D), part is reflected back, is intercepted by the drop, which, acting as a lens, returns the light back towards the sun and our eyes. The process is not perfect because the water drop is a poor lens and suffers severe optical aberrations. Owing to these deficient image-forming properties of water drops,

the light does not undergo a deviation of precisely 180° but rather comes out in a distribution that is a few degrees across: the heiligenschein. Like most surfaces, grass scatters light in a diffuse manner but with a preference for the backwards direction. Virtually all surfaces have this tendency to scatter light backwards[21,22].

The brightness of the heiligenschein depends critically on the distance between the drop and the blade of grass. It is particularly bright when the reflecting surface coincides with the focus. Since the drop rides on tiny hairs from the grass, it never actually touches and wets the blade. As the British meteorologist Tricker[23] has shown, the heiligenschein ceases to exist when the diffuse reflector is removed from behind the drop. The glint off the drop's front surface does not do the job. He also postulates that the color of the heiligenschein matches the color of the surface behind the drop, and experiments bear this out. Normally it is perceived as white because

Fig. 4.12A Heiligenschein on bedewed grass with camera at eye position.

Fig. 4.12B Same as in Figure 4.12A with camera held to one side.

the individual drops appear point-like and bright enough to saturate the eye's color receptors. The same will be true of photographs unless we underexpose the picture, or purposely defocus it.

The heiligenschein can also be seen from aircraft, but in this case there are other sources of retro-reflection in the landscape which produce similar backscattered glows (see Section 1.5).

Certain trees and shrubs show a sylvanshine. It seems to differ slightly from the heiligenschein on grass in the sense that the leaves may not have little hairs to support the drops. The drops are nearly spherical because the surface is not wetable[24].

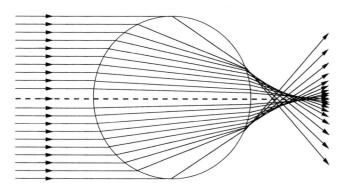

Fig. **4.12C** Ray paths through a water drop that focus light beyond it.

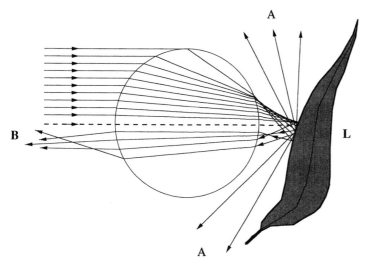

Fig. **4.12D** Same as in Figure 4.12C except with a leaf placed at best focus. B is the backscattered light that creates the heiligenschein; A, light that is lost; L, the leaf. The drop is supported by cuticle hairs, not shown.

4.13 Coronae

Colored concentric rings a few degrees across around the sun and moon when viewed through thin clouds are called coronae. They are easiest to see when the full moon is the light source (Figure 4.13A), provided the cloud is not too thick. Coronae colors go from white in the middle (an aureole) to blue, green, yellow, red and this series repeats outward for each successive ring. The most common coronae are found in altocumulus and consist simply of a bluish aureole plus a pale rust-colored ring surrounding it (Figure 4.13B). Coronae are unpolarized, which indicates reflections are not involved.

In daytime, coronae may pass unnoticed because we instinctively avoid looking towards the glare of the sun. To search for coronae, inspect the sky a few degrees from the sun while wearing dark sunglasses, or block the sun itself with your hand (Figure 4.13C). Another technique is to look at a reflection of the sun in a quiet puddle of water. Coronae are not rare; traces of coronae are found in many clouds.

Coronae are another of those phenomena whose explanation requires the wave theory of light. Coronae are produced by the interference of light diffracted around the outside of water drops or, in some regions and seasons, by pollen grains[25]. Although the drops are randomly spaced, it is the angle of diffraction that counts, and similar size drops act in unison to create what we perceive. The diffraction angle depends on both the drop diameter (Figure 4.13D) and the wavelength of light (Figure 4.13E). Larger drops cause smaller diffraction angles and a smaller corona. Red wavelengths constructively interfere at greater diffraction angles than blue. These two factors lead to great variability in coronae.

When for some reason the drop sizes are all about the same, the angle at which any color constructively interferes is fixed and the resulting color is relatively pure and bright. Under these circumstances the average size of the drops in the cloud may be deduced by measuring the angle between the sun and the inner-most red ring. Lenticular altocumulus, because of their short lifetimes, tend to have uniform drops, and these clouds display some of the best coronae. In thicker less transitory clouds the colors are usually muddy because the droplets have a wide range of sizes and the aureole then dominates.

Fig. 4.13A A regular corona around the moon. Star trails result from the time exposure.

Fig. 4.13B An aureole as blended with a poorly developed corona. The mix of color is dominated by browns with lesser amounts of metallic blue and green.

Fig. 4.13C Aureole plus regular corona around the sun.

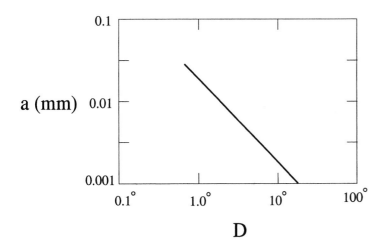

Fig. 4.13D A corona whose first (inner most) red ring is D degrees in diameter is formed by a drop of radius a.

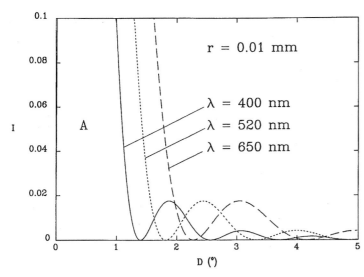

Fig. 4.13E Relation between corona ring intensity I and angle D from the sun for blue, green and red light of wavelength λ. A drop radius $r = 0.01$ millimeters is assumed. A is the region of an aureole.

4.14 Irisation

Color in clouds from corona fragments at large distances (say 5°–45°) from the sun is called 'irisation'. Splendid colors can be generated (Figure 4.14A). In common with the irisation of mother-of-pearl shells or morpho butterflies, iridescent clouds often take on the metallic hues indicative of overlapping orders (colors). Occasionally chromatic bands are found along certain cloud borders (Figure 4.14B).

It is obvious that iridescent effects within a few degrees of the sun are the same as coronae. But when a small cumulus cloud some 40° from the sun turns green, what corona order does this represent? Your authors have seen such an evenly colored small cumulus cloud many tens of degrees from the sun. As we explained, iridescent clouds tend to be purple and pink from the overlapping of orders of the diffraction spectra from different size particles. When red from the first order falls on blue from the second order, purple-pink results. A pure green color suggests a low order, since high orders would yield a metallic luster. It also implies a very uniform drop size. Assuming first order interference, a green cloud observed 45° from the sun implies 0.0012 millimeters droplets. This is possible, although on the extremely low side of the cloud droplet size distribution. The rarity of such clouds probably is a result of this stringent size and uniformity requirement, conditions that may be compatible with droplets being on the verge of formation or dissipation. Presumably cloud colors along their borders are also an indication of either uniform particles or a definite gradation in drop sizes with depth into the cloud.

Fig. **4.14A** Irisation with its metallic hues.

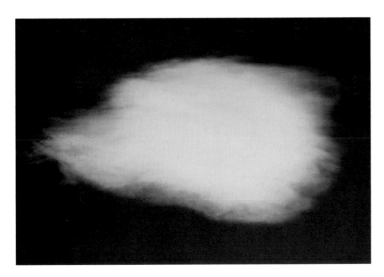

Fig. 4.14B Irisation in lenticular clouds. Along the border of the upper right cloud we see cloud banding with the edge rimmed consistently red to blue inward.

4.15 The glory

Standing on the edge of a fog-filled canyon with the sun at your back, you may see bright colorful rings around the antisolar point. If the fog is close by, your shadow is visible as a spectre centered on the rings. The rings have no set diameters but are less than $5°–10°$ degrees across. Similar to regular coronae, except centered on the antisolar point instead of the solar point, this glory is one of the beautiful apparitions of nature (Figures 4.15A, 4.10B).

A century ago the observation of a glory was considered a newsworthy event[26]. But today the glory is an everyday experience to you as an airline traveler (Figures 4.15B, 4.15C). Particularly when your plane passes between cumulus, or transits cloud decks, the glory is inevitably there to some degree provided the sun is favorably positioned at your back. It is most common to have one ring visible. When for some reason the drops in the clouds are especially uniform, the glory can display many rings. If your aircraft skirts a canyon of clouds, the ring diameters can fluctuate wildly with the aircraft's shadow coming and going to generate a dynamic spectacle.

Unlike coronae, glories are polarized[7], and this is a clue as to their origin; at least one reflection is involved. The sense of polarization is opposite to that of a rainbow, being radial to the rings. On the other hand, the bright central area is tangentially polarized except for the exact center which is unpolarized.

Fig. 4.15A Brocken spectre and glory observed at Puy du Dome, France. (Photo by V. Gaizauskas and C. Zwaan)

Fig. 4.15B Spectre and glory seen against a thin veil of mist over the sea. (Photo by R. Drake)

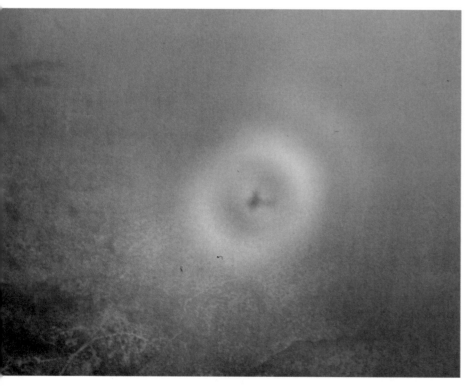

Fig. 4.15C A sequence repeated daily thousands of times world-wide. As the plane dips low it takes on an umbral shadow plus the glory. (Photo by O.R. Norton)

Although the glory pattern is correctly predicted by Mie theory[27–30], a good physical explanation is, in our opinion, lacking. In some way light is backscattered after traversing the periphery of droplet. Examined in detail, each drop is found to shine uniformly around its edge with an annulus of light that is coherent (the waves are in phase).

One of the best photographs of the glory was obtained[31] in 1968 from the rim of fog-filled Haleakala Crater on the Hawaiian island of Maui (Figure 4.15D). Five glory rings could be made out on the original transparency. Their radii measured 1.2°, 3.0°, 4.9°, 6.7° and 8.3° in agreement with the Mie calculations of H. C. van de Hulst[32]. Implied was a drop size of about 0.0018 millimeters.

There is some evidence that ice crystals in cirrus can produce glories[33]. Although ice crystals are not spherical, they can be very compact with their heights and widths about the same dimensions. These little crystals (0.01–0.02 millimeters) are not needle- or plate-like and have little in the way of extensions or projections. Since they have random orientations, their average cross section is roughly circular and this may produce faint glories. By the same reasoning, thin cirrus may occasionally produce coronae.

Fig. 4.15D High contrast reproduction from a Kodachrome of a glory taken on the rim of Haleakala Crater, Hawaii. Three, maybe four rings may be counted here. (Photo by J. Brandt)

4.16 Water drop optical effects

In the above sections we have attempted to explain the rainbow family employing the optical laws of refraction, reflection, and interference (or diffraction). Figure 4.16A brings together these diverse phenomena to emphasize how many ways sunlight and a little round drop of water can interact!

Besides the host of scattering phenomena, water drops can form actual images. Look closely at a water drop suspended from a leaf. You will see the hemisphere before you captured in amazing detail. Despite being upside down, the image is startlingly clear, with every element of the landscape rendered in complete clarity. Using a jewelers' loupe, you can see even more. The edge of the drop is sharply outlined in black. Focusing on different annuli from the center requires you to move the loupe in and out slightly, demonstrating that the focal plane is curved. But most obvious of all, the image is distorted with much of it being crowded up next to the edge. How does all this happen?

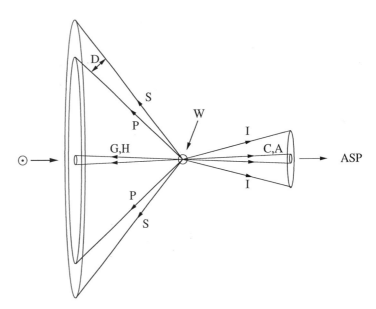

Fig. 4.16A Diagram of the water drop family of light scattering phenomena. W is the spherical water drop illuminated by the sun ⊙; ASP the antisolar point; A aureole; C coronae; D Alexander's dark band; G glory; H heiligenschein; P primary rainbow; S secondary rainbow; I iridescence.

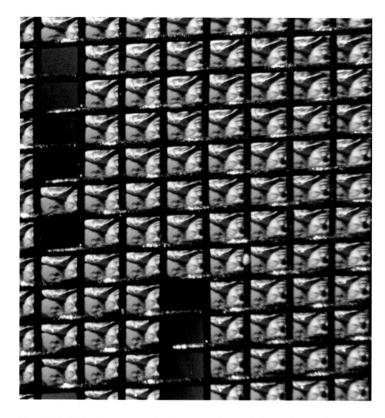

Fig. 4.16B Water drops on a window screen form miniature inverted images, in this case of a woman's face.

Water drops are natural 'fisheye' lenses, compressing a field of view of 165° into a nearly collimated beam (Figure 4.16C). They are roughly spherical, and the smaller they are, the closer to a true sphere. Even large distorted drops, such as one suspended from a leaf, have circular cross sections. The drop acts like a lens, bringing the image to a crude focus about one drop radius behind it. The image is upside down because you are viewing it outside the focal length of the real image. See the accompanying box on further experiments involving water drop images.

EXPERIMENTS WITH WATER LENSES

Find an old scrap of window screen (wire screen of 2–3 millimeter mesh) or, at any hardware store, purchase a screen repair kit. Dip a 4 centimeter square of screen into a bowl of water. Water will cling to the tiny screen openings and act as miniature lenses. Hold this screen up and notice that objects, such as a nearby tree, appear inverted. Examine these images with a magnifier (not shown). Determine the focal length of the lenses by imaging the sun onto a piece of paper. You will see an array of point-like solar images at a distance about one centimeter from the screen, the focal length. Since you can image the scene onto the paper, this is a *real* image.

With the passage of time, water begins to drain off the screen, or evaporate, and the screen will pop clear. Before the lenses clear, a strange thing happens, the lenses will cease to image the sun and only a diffuse smudge is seen on the paper. Directly examine the lenses and, behold(!), the image is now erect. These erect images cannot be imaged on paper and are referred to as *virtual*.

We have noticed that just before a lens pops, little dust-like hairs appear in the lenses. Maybe they are dust in the water that was not obvious before. Other things to try: honey or cooking oil, the effect of a wetting agent on water, or an anti-wetting material like oil. In principle one could make a *Galileo-type* telescope using the two types of above lenses (but your authors have not been successful in this).

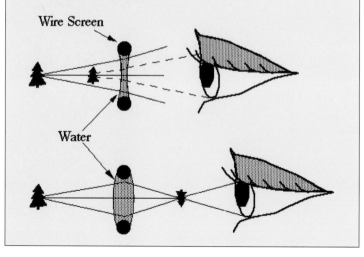

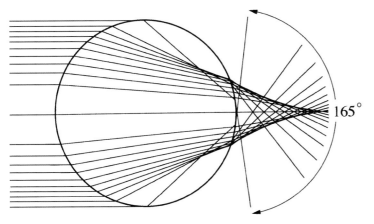

Fig. 4.16C Ray trace through a horizontal circular cross section of a pendant water drop. Note that almost the entire hemisphere is brought to a nearly collimated beam.

CLOUDY SKIES

4.17 What is a cloud?

We all know what clouds look like: they are those white, often well-defined bodies floating in the atmosphere. To the scientist, a cloud is a visible air-borne suspension of particles, usually water droplets or ice crystals. Their size varies from less than a micrometer (0.001 millimeters) up to about 100 micrometers[20]. A wide range of particle sizes is the rule. Drops larger than about 100 micrometers are not held in suspension but fall out as mist or rain. All cloud particles drift slowly downward; water drops about 1 micrometer across descend at only a fraction of a millimeter per second and so are lofted indefinitely at the whims of air currents. The 100 micrometer variety fall at about 30 centimeters per second.

Our ability to see a cloud depends on how it is illuminated, the size of its particles, and their number density (particles/cubic centimeter). Despite their impressive visibility, clouds are tenuous beings. A cumulus may have 1000 droplets per cubic centimeter, but these drops are minute and comparatively widely spaced. Water makes up less than one billionth (10^{-9}) of the cloud's apparent volume and contributes only about 1 millionth of a gram per cubic centimeter to its density.

What gives a cloud such definite form? Why are they not more diffuse in outline? The explanation involves the air's strictly limited ability to hold water vapor (an invisible gas), a state described by the relative humidity. At a given temperature the relative humidity is the amount of water vapor the air *is* holding divided by the amount it *can* hold, expressed as a percentage. When the relative humidity is equal to or greater than 100%, the air is said to be saturated or supersaturated and clouds materialize. Warm air has more capacity for water vapor than does cold. At sea level, saturated air at 25°C holds 23 grams of water vapor per cubic meter but at freezing (0°C) it can carry less than 5 grams per cubic meter. Under saturated conditions a water or ice cloud can persist with its particles either maintaining their size or growing. Inside a cloud the relative humidity exceeds 100%. Only at the edge, where the humidity may be less than 100%, can a water cloud dissipate by evaporation. Ice clouds disappear by sublimation, meaning solid water changes directly to vapor without first going through a liquid phase, and this is a slower process. Clouds have their distinct shapes because volumes of air tend to be lofted as discrete units. These parcels maintain their identities including their boundaries which are relatively sharp.

4.18 Why are clouds white?

Few objects in nature are white, yet clouds almost define the color. Why? The simple answer is that they are white because practically no light is absorbed and all colors are scattered equally. Usually illumination is from the sun, a source that is white to our eyes. Individual cloud particles are spherical and, if examined separately, might indeed show colors at certain sighting angles. In a cloud, however, the particles are too small and too mixed in size to show their individual wavelength signatures, so that colors run together in equal measure and appear white. This is especially true in thick clouds where the light may enter and bounce around many times before exiting. Although water itself has a faint bluish color in transmission, this hue plays no role in a cloud because light traverses such relatively small distances in traveling through individual drops or ice crystals. How light is scattered by jagged or irregular ice crystals is more complicated than for spherical water drops, but the end result is the same: ice clouds are white too.

4.19 Why are some clouds dark?

Examine a field of broken cumulus. Some clouds are bright, others are relatively dark (Figures 4.19A, 4.19B). While vigorously growing, a thunderhead is brilliant. As the cloud matures and ceases to soar, the upper part, or crown, perceptibly dims (Figure 4.19C). How can we account for these and other differences in cloud brightness?

Two factors are largely responsible for the variation in cloud brightnesses: (1) shadows and (2) cloud thickness or transparency[34]. For thick clouds where multiple scattering is important and where no background light shines directly through the cloud, the cloud's brightness is determined primarily by how much light falls on it. In full sunlight it is intensely white. In the shadow of another cloud, it appears gray in comparison. Dark clouds are not made of dirty water; they are simply dark in relation to the surrounding, more brightly lit clouds.

A thin cloud, on the other hand, will transmit some background light of the blue sky and much of the direct sunlight that passes through is lost to our eyes. Since nothing in the landscape is as bright as a white cloud, transmitted background light is dimmer and so the thin cloud appears darker.

An exception to this relation between thick and thin clouds happens when the cloud is back lit. Here a thick cloud appears dark because of self-shadowing. We see only its darkened underparts where very little light penetrates. A thin cloud, by comparison, is composed of small particles which scatter light forwards in a very efficient manner. Light transiting a tenuous cloud is only scattered a single time before escaping and the net result is a dazzling glare.

It is said that cumulus clouds darken with age[35]. This is reasonable since a new cloud will contain smaller particles. With time cloud particles grow by collisions with each other, and also the resulting larger drops lose water less readily by evaporation (since the internal pressure of small drops is greater from surface tension). Now, consider two clouds containing the same amount of water per cubic centimeter, but the older one has particles that are

Fig. 4.19A Front illuminated cumulus congestus. Note the smaller cloud (cumulus mediocris) which is relatively dark because it is thinner and a smaller fraction of the incident light is returned (multiple scattered) back to the front surface.

50 micrometers in diameter, while droplets in the newer clouds are only 5 micrometers across. The larger drops hold 1000 times more water than the tiny droplets, so there are 1000 times fewer drops per cubic centimeter than there are in the new cloud. Because the cross-section of a drop (and thus its ability to scatter light) is proportional to the square of its diameter, a 5 micrometer drop is only about 1% as efficient at scattering light than a 50 micrometer drop. But since there are a thousand times as many small ones as there are large ones, the scattering efficiency per gram of water is ten times greater for 5 micrometer drops than for 50 micrometer drops. This helps to explain why a cumulus cloud

Fig. 4.19B Clouds that are dark because they lie in shadow.

composed of small drops can be much brighter than low-level fog and stratus which contain larger drops.

As the top of a thunderhead pushes into the low stratosphere its drops either evaporate in the dry air there or freeze. Diminishing number density by evaporation darkens a cloud that continues to persist, now populated by ice crystals. This accounts for the diming of a cumulus anvil[36].

Fig. 4.19C In the foreground a growing cumulus congestus with bright upper surface. To the rear and above, a mature and now dissipating cumulus (with anvil) exhibits a darker crown because of larger (fewer) drops, while the outer rim is darker owing to transparency.

4.20 Colors of clouds

A low sun often leads to vivid cloud colors (Figure 4.20A). Earlier the sky may be completely overcast and dull. Now the sun emerges near the horizon and reddened sunlight pours through. Our day is suddenly made glorious by a palette of poly-chrome clouds. Practically every sunset has display potential.

Clouds being intrinsically white, any coloration must be external in cause: (1) they may simply be illuminated by colored light; (2) they may be translucent and allow colored background light through; or (3) they may be seen at a long distance through air which absorbs certain wavelengths and adds others.

Fig. 4.20A Cloud colors at sunset: in the foreground cumulus fractus, in back mammatus.

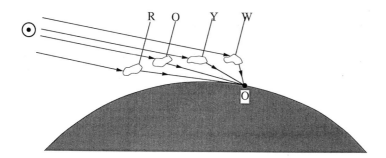

Fig. 4.20B Overhead clouds are illuminated by white light from the sun. They appear white (W). The closer these clouds are to the horizon of the observer O, the more reddened (Y yellow, O orange, R red) they are by scattering.

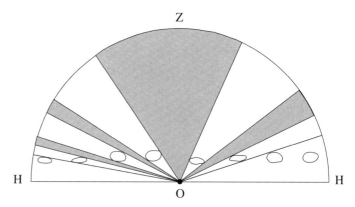

Fig. 4.21 Cloud blocking. A single layer of scattered clouds tends to obscure the horizon sky more than it does the overhead sky. This is because a nearly horizontal line-of-sight is more likely to reach a cloud than a vertical one.

Clouds glow pink, yellow, orange, or red at sunset because the sunlight falling on them has been reddened by atmospheric absorption and scattering. But there are other more subtle origins of color. A cloud over desert often looks slightly brownish as it reflects light from the brownish earth below (see Section 4.22). On clear days with a high sun, distant clouds on the horizon may appear either slightly pinkish or bluish (Figure 4.20B). Pink results from white light that has passed through so much atmosphere that it is reddened in the same way as the low sun, plus a weak veil of blue from intervening airlight. Blue clouds are found in the shadows of other clouds (see shadow colors) because they receive radiation from the bluish sky and not the white sun.

Over a city at night low-lying clouds may take on a yellowish hue. This coloration is particularly noticeable in the presence of higher level altocumulus or cirrus which then seem relatively colorless. We can suppose this yellow cast arises from the combined contribution of incandescent lights and sodium vapor street lamps.

4.21 Cloud blocking

When small clouds are scattered more-or-less randomly over the sky, they seem to be most numerous and crowded near the horizon, an effect called cloud blocking. Cloud blocking can lead to a relatively clear overhead sky but a totally obscured sky near the horizon. Cloud blocking occurs because clouds are found only in the relatively thin troposphere and often are confined to a single horizontal band (Figure 4.21). As our line-of-sight goes from zenith to horizon, it becomes more nearly parallel to the layer of clouds. Thus our line-of-sight intersects many more clouds near the horizon than it does overhead.

4.22 Blinks

Arctic travelers have learned that the brightness and hue of the underside of a low cloud layer reflects the surface conditions below and beyond it[37]. For obscure reasons these illuminations are called blinks. Clouds over water (Figure 4.22A) tend to be dark, giving rise to the 'water sky blink'. Cloud bases over ice are relatively bright, the 'ice blink', or over snow (Figure 4.22B), the 'snow blink'[37]. Similarly, dry land has its own cloud signatures: the 'land sky'. M. Minnaert[38] also noticed that clouds take on purplish tones over a heath-land if the heather is in bloom. In the southwestern United States the white sands near Almagorordo, New Mexico, produce a blink when that desert area is covered by open stratocumulus. Clouds that span a variety of surface conditions may display an induced brightness pattern called a 'sky map'. Such a map might conceivably be useful to an explorer when searching for a body of water.

Fig. 4.22A Water sky blink, as shown by the darkening of the cloud layer to the left near the horizon. (British Antarctic Survey, photo by J. Shankin)

Fig. 4.22B Snow blink, as shown by brightening of clouds to right near horizon. (British Antarctic Survey, photo by J. Shankin)

4.23 Does every cloud have a silver lining?

When a cumulus is back lit, its body is dark because the cloud is thick, but its periphery shines brightly (Figure 4.23). This is the silver lining. Seen thus, all cumulus-type clouds display this brilliant rim. The explanation is simple: around its edge a cloud is thin and light is efficiently scattered forward. Contrast this with a scene (Figure 4.19A) where the sun is behind us and we experience just the opposite: the edge of the cloud is dark and the center bright. Background light that penetrates the cloud is feeble and therefore the lining is dark by contrast to its front-illuminated body. The answer to the question is therefore no; it depends on your viewpoint.

Fig. 4.23 Late afternoon cumulus congestus back-illuminated to create a silver lining. Compare with Figure 4.19A which is front lit.

4.24 Walking in a fog

Inside a fog our visual world can collapse to distance of a few meters. All fogs are not equal and how far we can see ranges from 10 meters to 1000 meters, depending on the fog's water content[20] (Figure 4.24A). From its exterior a fog patch may appear puny (Figure 4.24B) or impressive (Figure 4.24C).

Most fog is composed of water drops but in cold regions ice fogs are commonplace. At the south pole the presence of floating ice crystals is almost the rule. While not a conventional fog, it is a cloud at ground level, and its crystals continually make weak halos.

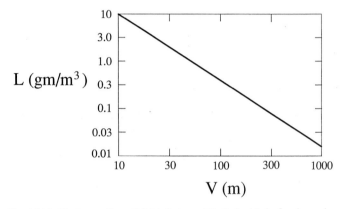

Fig. 4.24A Maximum line-of-sight distance (V) attainable in fog depends on the amount of suspended water per unit volume (L).

Fig. 4.24B Radiation fog fills valleys in the Appalachian Mountains. Such fog patches are likely places for spectre shadows.

Fig. 4.24C Advection fog fills the Los Angeles basin to the brim. Here it spills over the shoulder of Mt Harvard into a canyon on the side of Mt Wilson in Southern California. On a moonless night lights from the valley below cause this fog to emit a yellowish glow.

4.25 Visibility of the sun through a cloud

When the sun shines through low clouds, its disk is sharp and may be easy to look at without discomfort. No obvious aureole surrounds it. Yet when viewed through high clouds like thin altostratus, the disk seems fuzzy and indistinct. The edge, if visible at all, is vague and there is a strong aureole. The presence of the aureole seem to be important in lowering the apparent contrast of the limb of the sun. The optical thickness of the cloud is obviously important since it must be large enough to attenuate the sun's light to the point of comfortable viewing. But what exactly causes fuzzy suns?

No one is certain although the answer may involve a special combination of particle size and cloud thickness[39]. Experiments in the laboratory suggest that any size of cloud particle can probably produce a fuzzy sun with a strong aureole but only when its optical thickness is within a certain limited range. Since clouds are continuously changing, the chances of seeing a fuzzy sun though the cloud depend on looking when the particle size and optical thickness are just right. In general, the larger the particle size, the wider the range of cloud thicknesses that can cause the sun to appear fuzzy. The optimum optical depth seems to be in the range 5–10.

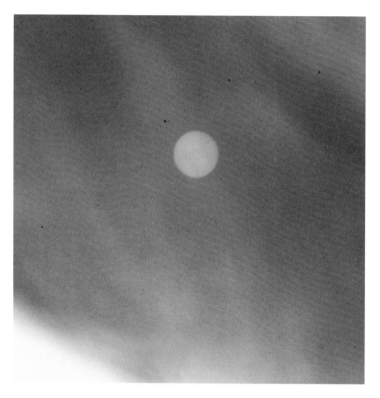

Fig. 4.26A Rare blue sun as photographed at Boulder, Colorado, by P. Neiman.

4.26 Once in a blue moon

On rare occasions the moon (sun) looks blue, or green, or purple (Figure 4.26A). This only happens when the moon is seen through a cloud, haze, or smoke[40-2]. Such a sight will happen 'once in a blue moon'.

The reason for a blue moon involves that no man's land where the wavelength of light is comparable to the particle size (0.1–10.0 micrometers). In this transition region, the scattering efficiency can vary strongly throughout the visible spectrum (Figure 4.26B). Such scattering acts like a filter to sunlight: it transmits some colors and scatters other colors out of the beam. These effects are suppressed by the range of particle sizes ordinarily present in haze or clouds. But when the range is narrow and the size just right they act in concert to create a blue moon[32, 43].

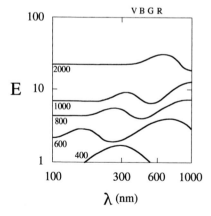

Fig. 4.26B Optical basis for the blue sun/moon. The scattering efficiency E of drops of various radii shown as a function of wavelength λ. Notice that drops with a radius of 400 nanometers scatter more efficiently in the blue and therefore the transmitted solar (or lunar) image would be reddish. Drops with radii of 600 nanometers scatter better at longer wavelengths and therefore the transmitted image would be blue-green.

4.27 Haze, smog, and smoke

We are in the orient. It is not cloudy but there is no sun. There seldom is, in fact, for this quarter of the globe lives under a perpetual haze layer. In India it is called the *godalli*, meaning cow dust in Sanscrit. It extends from Bombay to Calcutta to Hong Kong: an all pervasive ground-hugging haze made up of cooking smoke, water vapor haze, and dust. At the horizon the solar disk is virtually extinguished. As the sun rises its reddened image ever so reluctantly materializes. Even at noon the sky is milky. Many of the atmospheric optical displays described in this book are unobservable in this part of the world because the sky is never clear.

Elsewhere there is smog. This is the product of automobile exhaust and other pollutants containing nitrogen dioxide as well as carbon and sulfur particles which lead to the formation of numerous aerosols. Smog is photochemical. Sunlight catalyzes the reactions, as do rising temperatures. Smog is worse at mid-afternoon and dims the low sun.

4.28 Contrails and distrails

Alternatively named contrails or vapor trails, these are the condensation clouds that form behind aircraft. High level contrails can be the site of vivid halos. This is because their newly created ice crystals tend to be uniform in size and shape.

Two related processes can account for contrails. First, water vapor from engine exhaust may condense. This condition is especially favored for high flying jets near the cold tropopause where even a small addition of moisture prompts immediate cloud formation. Second, exhaust particles ('soot') may act as condensation nuclei for existing water vapor.

At lower elevations and particularly in warm moist air, the aerodynamic pressure reduction behind an airfoil will induce condensation. Such contrails commonly emerge from wing and propeller tips. We have noticed that a DC 10 is particularly prone to aerodynamic condensation over the wings on take off and landing. When the leading edge flaps are extended a tubular pattern loops over the wing. We have seen this streaming condensation continue to an elevation of 11 000 meters.

Occasionally, instead of the usual continuous condensation, a series of 'smoke ring-like' patterns is laid down. These smoke rings, no doubt wafted by local air currents, immediately distort into figure eights and other curious formations (Figure 4.28). We have no clue as to why this happens. Within our experience the sightings (four in number including one from the literature[44]) have all been over the British Isles. See also p. 122 in Scorer[35].

The shadow of our plane's contrail can cause a dark line to emerge from any accompanying glory. Other contrail shadows may crisscross the cloudscape (Section 1.4).

When a jet flies through a thin cloud it can be so disruptive that a section of the cloud disappears. This is a distrail (DISsipation TRAIL) and is the opposite of a contrail.

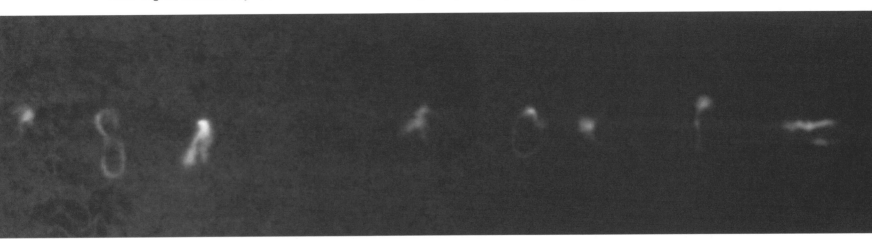

Fig. 4.28 Ring-like patterns in a contrail from a jet over Ireland.

ELECTRICITY

4.29 Lightning

Lightning should be enjoyed from a great distance! Estimate the distance to a flash by figuring 0.34 kilometers per second for the delay time of the thunder. On the other hand, thunder is seldom heard over 25 kilometers (delay over 1 minute) and often it is imperceptible at half that distance.

A typical lightning flash lasts a fraction of a second[45]. Its flicker tells us the discharge consists of several shorter duration strokes. In the presence of a wind the lightning channel may be blown sideways to produce ribbon lightning from these multiple strokes. By means of photography at night we can resolve the individual strokes if we swing a hand-held camera back and forth over an arc of about 20°. The resulting picture, if successful, will show that the first stroke is usually the brightest[46]. Some strokes appear broad and faint, indicating the condition of a 'continuing current'. While all lightning strokes can be destructive, it is those with a continuing current that start fires. A flash can involve 25 or more strokes, each of a tenth of a microsecond in duration.

Lightning is a giant electrical spark that attempts to neutralize the potential difference in regions of positive and negative charges. One theory suggests that falling raindrops become distorted and electrically polarized. If the drop breaks up, the lower, larger part will tend to be left positive; the upper, smaller part will be negative (Lenard effect). Now let strong vertical wind currents carry the finer drops toward the upper reaches of the cloud, while the big drops descend under gravity. Finally the top of the cloud is charged negative, the bottom positive. In this case the earth under such a cloud will assume a negative charge[47].

Discharges between cloud and earth are called 'ground flashes'[46]. Discharges within or between clouds are 'intracloud lightning' and these are many times more frequent than the ground flashes. A lesser event is 'St Elmo's fire', the continuous brush discharge from elevated pointed objects such as church spires or mountain summits (Andes lights).

Why the lightning channel follows its characteristic zigzag route in such an unpredictable way is not understood[48]. Kinks as small as a few centimeters are found from telephoto pictures[49]. Near the ground, one or more strokes will frequently depart from the initial discharge channel to cause dendritic or 'forked lightning' (Figure 4.29A). Less common is the dissolution, or breakup, of a stroke; persistent elements give rise to the term beaded lightning[49]. Its cause is also unknown.

Then there is the mysterious ball lightning[50,51]. Some kind of luminous sphere, 2–100 centimeters in diameter, is generated as an offspring of a lightning flash. For several seconds this ball may move in a capricious manner, slowly drifting through the air. Most astonishing, the ball has been seen to go against air currents and through window panes without their breakage. There has been little agreement as to cause or what ball lightning is, although some ideas evoke the concept of a special state of ionized gas, a plasma. Kapitza, the famous Russian physicist, suggested that radiofrequency discharges may be created by some kind of natural condition. Recent reports have it that a thunderstorm is not even a requirement[52].

Volcanic eruptions foster another bizarre form of lightning[49]. The convective ash cloud becomes charged and witnesses report an almost continual discharging into the surrounding clear air.

Lightning displays are easily photographed at night by making a time exposure of a few minutes depending on the brightness of your surroundings. Flashes may serve to 'strobe' the growth of cumulus clouds so that our time exposure becomes a stop-action movie[53] (Figure 4.29B). Many published lightning photographs show this strobing effect although it is never mentioned. Look at cloud tops photographs showing several flashes and notice how often adjacent clouds have similar contours. The same cloud is being repeatedly recorded by lightning as the cloud convects skyward.

Even though lightning itself is usually colorless, the intervening atmosphere can redden distant flashes and allow us, by color film, to judge the relative distances of multiple events (Figure 4.29C).

In daytime a flash can often be captured by pointing the camera in the likely direction, posing one's finger on the shutter button, and firing off a shot when a flash is seen[54]. Of course the stroke you see is not the one recorded, but rather the subsequent strokes, Figure 4.29E.

Fig. 4.29A Forked lightning, in which one of the later strokes seeks a different route to the ground.

From space, the night-time earth reveals many regions of almost continuous thunderstorm activity. Satellite observations inform us that a majority of such storms are over land and not the oceans.

4.30 Sprites and jets

For more than a century people have reported strange, diffuse flashes of light in the night sky over thunderstorms. These flashes have recently been studied from the vantage point of space and are now called sprites, jets, and elves because of their differing shapes and elusive nature[55,56]. They are usually red or blue and occur at a height between 45 and 85 kilometers. Sprites and jets are barely visible to the unaided eye because they are only as bright as a moderate aurora and last but a fraction of a second. Red sprites[57] are the most common, an event taking place simultaneously with conventional lightning underneath. Their shape tends to be diffuse and irregular, extending from the top of the thundercloud to the ionosphere. Blue jets[58] are more compact and may appear in clusters traveling upwards from the top of the cloud.

Sprites and jets are a newly recognized kind of lighting discharge. Their color is the result of narrow band emission from atmospheric atoms and molecules, especially excited nitrogen.

Fig. 4.29B Time exposure during a lightning storm causes multiple images of cloud tops as they rapidly grow.

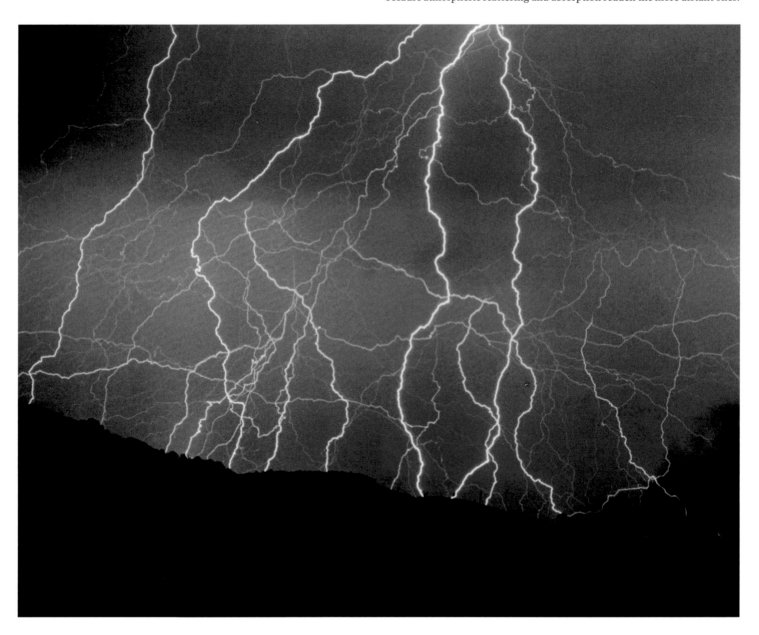

Fig. 4.29C Cloud-to-ground lightning flashes appear to have different colors because atmospheric scattering and absorption redden the more distant ones.

HOW TO TIME RESOLVE LIGHTNING

A lightning flash is not a single event but rather a sequence of strokes. We see this as flickering. These multiple strokes, though similar in that they follow the same channel, have subtle differences which reveal curious features of the discharge. For instance, in a time resolve picture the first stroke often has branches that are absent from the subsequent strokes. Continuing current shows up as a ribbon-like stroke. Some strokes produce knot-like bright points from which we see streaks. Some of these may be channels seen end-on, but not all.

It is easy to create a photographic time sequence with no special equipment. Just a camera with time or bulb exposure control, and patience, is required. This only works at night. When a lightning storm is underway, open the shutter and slowly swing the camera back and forth left–right about 45° in width. The time per swing might be 2 seconds. Keep doing this until a flash occurs within your field of view. If nothing happens, after perhaps 30 seconds, release your shutter, advance the film and start over. With any luck, in a given storm you may record many multiple flashes. The nice thing is that because each individual stroke is of a microsecond duration the moving camera images will be perfectly sharp. Also lens aperture is not critical. Every picture is different.

If a distant neighbor's porch lamp happens to be included in the scene this will produce a dotted line on the frame which can be useful for timing. The 60 hertz mains cause the light to fluctuate so that the traced dashes are 1/120 = 0.0083 seconds apart, and this means that the time interval between strokes can be measured as well as the total duration of the event!

Figure 4.29D is one result from such a procedure. We know time runs left to right because it is (almost always) the first stroke that branches. On the fifth stroke notice the bright point and that this coincides with an indication of continuing current on the last stroke. The bright point on the upper branch to the left is probably a projection effect. Timing light tracks (courtesy of our neighbor – also not reproduced here) allow us to state that the total duration of this event was 0.58 seconds.

Fig. 4.29D Time sequence of a lightning flash.

Fig. 4.29E Daytime capture of a lightning flash that includes the sun.

References

1 Boyer, C.B., 1987, *The Rainbow*, Princeton University Press, Princeton.

2 Gedzelman, S.D., 1982, 'Rainbow brightness', *Applied Optics*, **21**, 3032.

3 Floor, C., 1980, 'Rainbows and halos in lighthouse beams', *Weather*, **35**, 203.

4 Greenler, R., 1980, *Rainbows, Halos, and Glories*, Cambridge University Press, Cambridge.

5 Mattsson, J.O., Nordbeck, S., and Rystedt, B., 1971, 'Dewbows and fogbows in divergent light', *Lund Studies in Geography Series C, No. 11*.

6 Scott, G.D., 1975, 'The swimmer's twin rainbows', *American Journal of Physics*, **43**, 460.

7 Können, G.P., 1985, *Polarized Light in Nature*, Cambridge University Press, Cambridge.

8 Fraser, A.B., 1983, 'Chasing rainbows', *Weatherwise*, **36**, 280.

9 Lynch, D.K. and Schwartz, P., 1991, 'Rainbows and fogbows', *Applied Optics*, **30**, 3415 –3420.

10 McDonald, J.E., 1954, 'The shape of raindrops', *Scientific American*, **190**, 64 (Feb).

11 McDonald, J.E., 1954, 'The shape and aerodynamics of large raindrops', *Journal of Meteorology*, **11**, 478.

12 Volz, F., 1961, Chap. 14, Sec. 3 'Der regenbogen' in *Handbuch der Geophysik*, Linke, F. and Moller, F., Eds., Gebruder Borntraeger, Berlin.

13 Green, A.W., 1975, 'An approximation for the shape of large raindrops', *Journal of Applied Meteorology*, **14**, 1578.

14 Walker, J.D., 1976, 'Multiple rainbows from single drops of water and other liquids', *American Journal of Physics*, **44**, 421.

15 Lock, J.A., 1987, 'Theory of the observations of high-order rainbows from a single water drop', *Applied Optics*, **26**, 5291.

16 Walker, J.D., 1978, in 'The amateur scientist', *Scientific American*, **239**, 185 –187.

17 Boyer, C.B., 1959, *The Rainbow*, Thomas Yoseloff, New York; reprinted 1987, Princeton University Press, Princeton.

18 Pilsbury, R.K., 1991, 'Double white rainbow', *Weather*, **46**, 53–54.

19 Livingston, W.C., 1979, 'The cloud contrast bow as seen from high flying aircraft', *Weather*, **34**, 16.

20 McCartney, E.J., 1976, *Optics of the Atmosphere*, J. Wiley and Sons, New York.

21 Trowbridge, T.S., 1978, 'Retroreflection from rough surfaces', *Journal of the Optical Society of America*, **68**, 1225.

22 Trowbridge, T.S., 1984, 'Rough-surface retroreflection by focussing and shadowing below a randomly undulating interface', *Journal of the Optical Society of America*, A/**1**, 1019.

23 Tricker, R.A.R., 1970, *Introduction to Meteorological Optics*, American Elsevier Pub. Co., New York.

24 Fraser, A.B., 1994, The sylvanshine: retroreflection from dew-covered trees, *Applied Optics*, **33**, 4539.

25 Traenkle, E. and Mielke, B., 1994, 'Simulation and analysis of pollen coronas', *Applied Optics*, **33**, 4552.

26 Flammarion, C., 1874, *The Atmosphere*, Harper and Bros., New York.

27 Bryant, H.C. and Cox, J.A., 1966, 'Mie theory and the glory', *Journal of the Optical Society of America,* **56**, 1529 (July).

28 Bryant, H.C. and Jarmie, N., (July) 1974, 'The glory', *Scientific American*, **231**, 60.

29 Khare, V. and Nussenzweig, H.M., 1977, 'Theory of the glory', *Physical Review Letters*, **38**, 1279.

30 Nussenzweig, H.M., 1979, 'Complex angular momentum theory of the rainbow and glory', *Journal of the Optical Society of America*, **69**, 1068 (photo p.1194).

31 Brandt, J.C., 1968, 'An unusual observation of the 'Glory', *Publications of the Astronomical Society of the Pacific*, **80**, 25.

32 Hulst, H.C. van de, 1957, *Light Scattering by Small particles*, Wiley, New York.

33 Sassen, K., Arnott, W.P., Barnett, J.M. and Aulenbach, S., 1998, 'Can cirrus clouds produce glories?', *Applied Optics*, **37**, (9) 1427–1433.

34 Bohren, C.F., 1986, 'Black clouds', *Weatherwise*, **39**, 169; also **39**, 271.

35 Scorer, R.S., 1990, *Satellite as Microscope*, Ellis Horwood, New York.

36 Scorer, R.S., 1972, *Clouds of the World*, Stackpole Books, Harisburg, Pa.

37 Middleton, W.E.K., 1954, 'The color of the overcast sky', *Journal of the Optical Society of America*, **44**, 793–798.

38 Minnaert, M., 1954, *Light and Colour in the Open Air*, Dover, New York.

39 Linskens, J.R. and Bohren, C.F., 1994, 'Appearance of the Sun and Moon seen through clouds', *Applied Optics*, **33**, 4733.

40 Paul, W. and Jones, R.V., 1951, 'Blue sun and moon', *Nature*, **168**, 554.

41 Lothian, G.F., 1951, 'Blue sun and moon', *Nature*, **168**, 1086.

42 Green, H.L., Lane, W.R., and Hartley, H., 1964, *Particulate Clouds: Dusts, Smokes, and Mists*, Van Nostrand, New Jersey.

43 Wilson, R., 1952, 'The blue sun of 1950 September', *Occasional Notes, Royal Astronomical Society*, **2**, 137.

44 Allaby, M., 1995, p.116 in *How the Weather Works*, Dorling Kindersley, London.

45 Uman, M.A., 1969, *Lightning*, McGraw-Hill, New York.

46 Hendry Jr, J. and Uman, M.A., 1993, 'Panning for lightning', *Weatherwise*, **45**, 18.

47 Williams, E.R., (November) 1988, 'The electrification of thunderstorms', *Scientific American*, **259**, 88.

48 Krider, E.P. and Alejandro, S.B., 1983, 'Lightning, an unusual case study', *Weatherwise*, **36**, 71.

49 Salanave, L.E., 1980, *Lightning and Its Spectrum*, University Arizona Press, Tucson.

50 Ritchie, D.J., 1961, *Ball Lightning – a Collection of Soviet Research in English Translation*, Consultants Bureau, New York.

51 Singer, S., 1971, *The Nature of Ball Lightning*, Plenum Press, New York.

52 Singer, S., 1991, 'Great balls of fire', *Nature*, **350**, 108–109.

53 Livingston, W., 1984, 'Strobing cumulus growth by means of lightning', *Weather*, **39**, 240.

54 Beasley, W., 1982, 'Photograph of lightning in the daytime', *EOS Transactions of American Geophysical Union*, **63**, 802.

55 Lyons, W.A., 1994, 'Characteristics of luminous structures in the stratosphere above thunderstorms as imaged by low-light video', *Geophysical Research Letters*, **21**, 875–878.

56 Mende, S.B., Sentman, D.D. and Wescott, E.M., 1997, Lightning between earth and space, *Scientific American*, **277**, 56 (Aug).

57 Sentman, D.D., Wescott, E.M., Osborne, D.L., Hampton, D.L. and Heavner, M.J., 1995, 'Preliminary results from the Sprites94 Aircraft Campaign, 1: Red sprites', *Geophysical Research Letters*, **22** (10) 1205.

58 Wescott, E.M., Sentman, D., Osborne, D., Hampton, D. and Heavner, M., 1995, Preliminary results from the Sprites94 Aircraft Campaign, 2: Blue jets, *Geophysical Research Letters*, **22** (10) 1209.

5 Ice and halos

I will always remember my first sun pillar. I was walking to high school early one morning through a blanket of freshly fallen snow. Crossing a field I saw a vertical shaft of yellow-orange-light sticking straight up over the rising sun. It was sharp edged and absolutely straight, as though drawn by a draftsman. I hurried to school and told everyone about it. Had they seen it? I asked my science teacher about it. What was it? My excited story was met with curious stares and awkward silence.

Winter of 1963–4, South Bend, Indiana

5.1 Glints and sparkles

Snow is white because it absorbs very little visible light and scatters light very efficiently. With a reflectivity of about 95%, it is one of the most highly reflective substances known[1]. But a close look at snow reveals more.

Sunlit snow is sprinkled with points of light that are much brighter than the average snow. Some of them are white, others are colored (Figure 5.1A). To distinguish them we call the white ones glints and the colored ones sparkles. While glints may be found at any angle relative to the sun, sparkles tend to concentrate in a band roughly 22° from the sun. Their colors range over the entire spectrum. Glints and sparkles are also seen in frost.

Glints are caused by specular reflection off crystal faces. Sparkles are due to prismatic dispersion through the same kinds of ice particles (Figure 5.1B). Owing to the random orientation of the crystals relative to the sun, the prism angles vary on either side of 60° resulting in a broad angular spread of sparkles.

Viewing sparkles at close range is tricky because the dispersed light is diverging from the crystals. If we focus our eyes on a single crystal and yet are close enough to it that all of its dispersed light enters our eye, we will see no color at all. Only by viewing the crystal with an out-of-focus gaze can we see its colors. An easier method, of course, is to stand back a little so that only part of the dispersed light enters our eye.

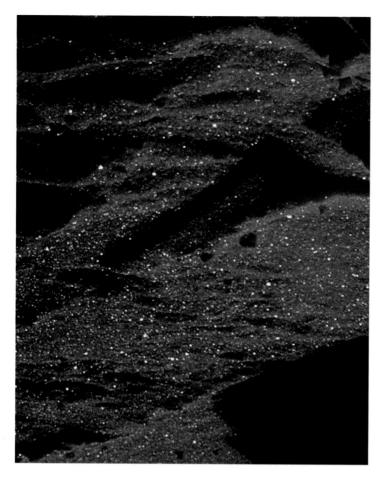

Fig. 5.1A Glints and sparkles are brilliant points of light seen most often in freshly fallen snow, hoar frost, and rime. Glints are colorless and sparkles show every color of the spectrum. (Boulder, Colorado)

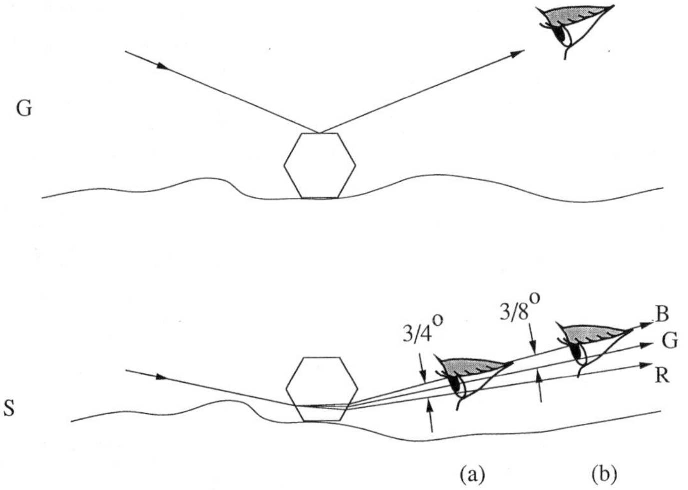

Fig. 5.1B Glints are simply specular reflections of sunlight from the surface of ice crystals (G). Sparkles are bits of sunlight dispersed by an ice prism (S). Nearby observers (a) may see less color than more distant observers (b) because their eye receives a larger range of colors which are brought back together on the retina as white. Only by viewing the sparkle with out-of-focus vision can they spread the colors on the retina and fully perceive them.

5.2 Ice and its optical properties

Ice is a hexagonal crystal. It can take on many forms[2–4], not just stellar snow flakes. In cirrus clouds ice tends to form plates (like old-fashioned bathroom tiles) or columns shaped like pencils (Figure 5.2A). Sometimes the plates and columns are terminated with hexagonal pyramids and other times they are too irregular to classify. During the life of the cloud the crystals may change size, shape, and optical properties. They melt, freeze, accumulate rime, or grow more complex shapes in response to the changing temperature and humidity[5–7]. Snow particles quickly lose their angular edges and metamor-phose into rounded nodules[8]. The nightly deposition of hoar frost tends to restore ice facets which regenerates glints and sparkles.

Like all substances, ice's index of refraction changes slightly with wavelength, being larger in the blue than in the red. This dispersion is responsible for the colors observed in ice crystal phenomena. Ice actually has two indices of refraction corresponding to its two principal axes, the a and c axes[9,10]. It is therefore called birefringent. The two indices, however, are so close together that for our purposes they may be considered as one (Figure 5.2B).

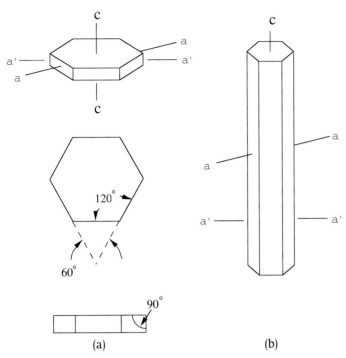

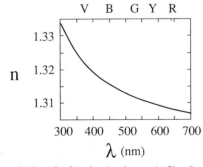

(a)   (b)

Fig. 5.2A Ice is a right hexagonal prism. In clouds, it frequently occurs as plates (a) and columns (b). The *c* axis passes perpendicularly through the center of the six sided basal face. The *a* axis is perpendicular to the *c* axis and passes through opposite 120° prism faces. The *a′* axis, which is also perpendicular to the *c* axis, passes through the 120° vertices. All axes meet in the center of the crystal. There is endless variation on the hexagonal ice crystal, including the pyramidally terminated bullets and rosettes made of three or more bullets joined at their pyramidal terminations. Most ice particles are irregular lumps that do not show interesting optical effects.

Fig. 5.2B Average index of refraction (real part *n*) of ice. Ice is birefringent and has two indices of refraction corresponding to the *a* and *c* axes. The indices are so close together that for most purposes, only one need be used for a given wavelength.

5.3 Color in snow banks and glaciers

Poke a deep hole in snow and what do you see? A blue-green light[11,12]. Does this remind you of glaciers (Figure 5.3A). If snow and ice are white, then why is light filtering through them blue?

Like water, ice absorbs a bit more in the red than in the blue (Figure 5.3B). Light emerging from a snow hole or a glacier has undergone many passages through the snow particles and the cumulative effects of absorption become evident. The initially white light gradually fades to blue. The deeper the ice, the deeper the color.

Sometimes icebergs look green, rather than blue. This may be due to a combination of low-sun reddish light mixing with the filtered blue ice light to make green[13]. Icebergs also contain more than ice. They have suspended sediments, algae, and air bubbles.

Fig. 5.3A When seen by reflected light, ice (and snow) is white. But when light diffuses through it as in a glacier, ice's intrinsic blue color is revealed. Both reflected and filtered light can be seen at the mouth of this ice cave. (Grindelwald, Switzerland)

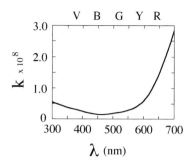

Fig. 5.3B The imaginary part of the index of refraction, k, is a measure of the absorption of ice at each wavelength. While refraction is controlled by the real part of the index of refraction, absorption is determined by the imaginary part. It reaches a minimum in the blue-green where ice is most transparent ($k \simeq 2 \times 10^{-9}$), and absorbs heavily in the red.

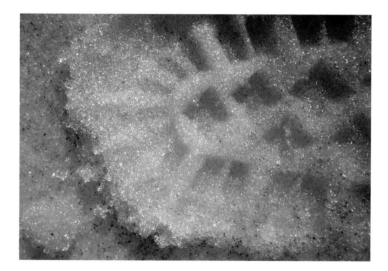

Fig. 5.3C Snow fields in the high mountains often show reddish areas. The source of the color is the algae of genus *Chlamydomonas*, which thrives in snow[14]. Mountaineers' footprints help to increase algae up to 500 000 organisms per milliliter and make them visible.

5.4 Introduction to halos

Halos are bright, often colorful spots, rings, and arcs seen in cirrus clouds and ice fogs[15-20]. The word halo conjures up the image of a circle, but few halos are actually circular. Most describe complex curves. They are formed by reflection and refraction through ice crystals. Halos occur frequently the world over. This is because the temperature in the upper troposphere is cold enough to freeze water, −40 °C or colder. Southern California is typical. In Los Angeles in any week several different halos can be seen by the alert observer.

The colors most visible in halos are the longer wavelengths: red, orange, and yellow. There are three reasons for this. First, dispersion is least for these wavelengths and they are not spread over nearly as large an angle in the sky as the short wavelengths are. This makes them brighter than the short wavelengths. Second, most halos are seen against the bluish background sky. The greens and blues blend in, leaving the reds and oranges showing the strongest color contrast. Third, when the sun is low in the sky, it usually appears yellow-orange in color due to Rayleigh scattering and absorption. The ice crystals receive more of the longer wavelengths than the shorter ones, and scatter them accordingly.

Despite their numbers and complex shapes[21-23], halos can be classified according to the particular prisms or crystal faces that produce them. Most halos fall into two categories, those associated with the 60° prism and those formed by the 90° prism (Figure 5.4A).

Crystals that produce interesting optical phenomena are almost always plates and columns. They are identical in shape except for their length-to-width ratio. Each crystal has three different prisms on it, one for each angle made by different faces (60°, 90°, and 120°). Only two are important: the 60° prisms between alternate faces and the 90° prisms between side and end faces. The 120° prism between adjacent faces plays only a minor and somewhat subtle role in refraction because total internal reflection prevents any light entering the first face from leaving the second.

By calculating how light passes through a prism, we can learn a great deal about halos (Figure 5.4B). Starting with a ray that is perpendicular (or normal) to the entrance face, we increase the angle of incidence (*i*) and compute where the ray goes and at what angle it will strike the second face before leaving the crystal. We then

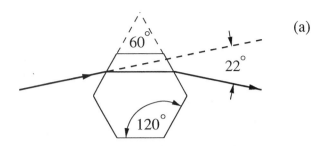

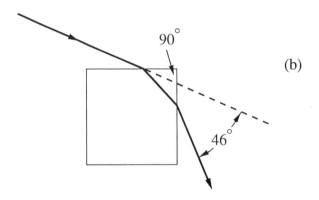

Fig. 5.4A There are three prism angles in the ice crystal: 60°, 90°, and 120°. Only the 60° (a) and 90° (b) prisms produce prismatic colors because light is totally reflected internally by the 120° prism.

calculate the angle through which refraction at the entrance and exit faces alters its path. This is called the deviation angle *D*. We find that the deviation reaches a minimum where the light passes symmetrically through the prism and is larger for all other paths. For light in the normal plane, i.e. the plane that is perpendicular to the refracting edge defined by the intersection of the two crystal faces, the angle of minimum deviation is about 22° for the 60° prism and 46° for the 90° prism. For rays that are not in the normal plane the angle of minimum deviation is always larger than 22° (46°) because the effective prism angle is larger. Like all reflection and refraction phenomena, halos are partially polarized[23].

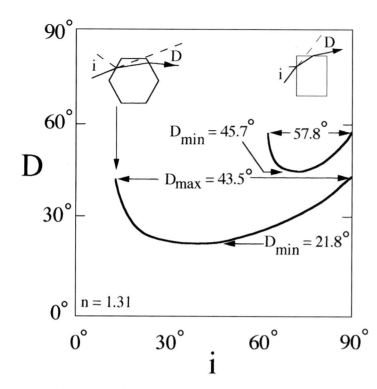

Fig. 5.4B For light passing through each prism perpendicular to the axis, the angle of minimum deviation D_{min} is about 22° for the 60° prism and 46° for the 90° prism. The angle of deviation is at a minimum when the light passes symmetrically through the prism and is greater at all other angles of incidence.

Table 5.4. *Orientations and crystal types.*

Orientation	c axis	a axis	Crystal type
{1}	V	H	plate
{2}	H	V	column
{3}	H	H	column

H = horizontal V = vertical

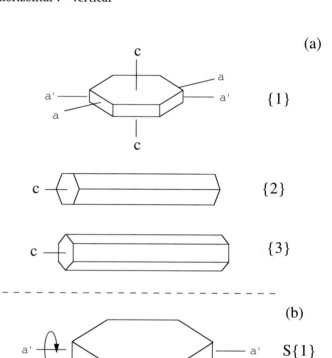

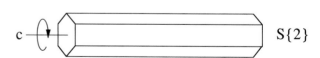

Fig. 5.4C Crystals become oriented by aerodynamic forces as they fall through the atmosphere (a). Plates descend with their c axes vertical {1}. Columns assume an orientation with their c-axes horizontal and either the a {2} or a′ {3} axis vertical. The orientation is often imperfect with significant departures from strict alignment. The very largest crystals spin about their horizontally-oriented a′ and c axes (b).

Like falling leaves, ice crystals may become preferentially oriented as they fall through the air[26,27] (Figure 5.4C(a)). Crystals smaller than about 0.025 millimeters across generally do not orient unless they are long and thin like needles or very thin like fish scales. Between about 0.025 and 0.25 millimeters crystals become partially or totally aligned in one of three possible orientations (Table 5.4). The degree to which crystals of any size become aligned depends on their aspect ratio, i.e. the ratio of their thickness (c axis) to width (a axis). Crystals whose longest axis exceeds about 0.025 millimeters probably orient themselves to some extent. Those whose width and length are nearly equal probably do not orient themselves regardless of their size. Halos formed by oriented crystals generally change shape with solar altitude.

Crystals whose long axis exceeds about 0.25 millimeters spin about this axis which is horizontally oriented (Figure 5.4C). Plates rotate about their a' axis and columns spin about their c axis. Can a plate ever rotate about its a axis, that is the one intermediate in length between the longest and shortest? No, at least not with any stability. Try spinning this book about its axes: the motion is stable around the long and short axes but not the intermediate one.

Halos formed in spinning crystals are the most complex of all because their size and shape change enormously with solar altitude. They can be classified into four groups, corresponding to the possible combinations of prism angle (60° or 90°) and location relative to the 22° and 46° halos (lateral, or on either side of the halo, and tangent, or above and below the halo). Since the 60° and 90° prisms produce the 22° and 46° halos, the halos due to spinning crystals are always found touching these halos, or at least touching the place in the sky where the circular halo would be if it was present.

The range of crystals sizes in a cirrus cloud is often large enough for randomly oriented, vertically oriented and spinning crystals to occur together. When this happens, many different halos may be seen at the same time. These are called halo displays.

In the following sections we discuss the halos in some detail. Figure 5.4D shows a schematic drawing of the most common halos drawn to show their approximate positions. Owing to the large number of halos, we will describe two in greater detail than the others: the 22° halo and the parhelion. They illustrate the fundamental optical processes that are common to many halos.

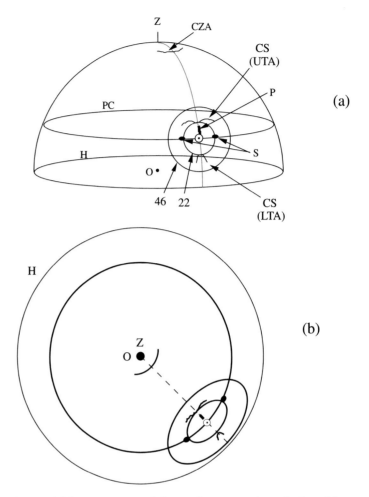

Fig. 5.4D (a) The most common halos are shown on a schematic view of the celestial sphere. The horizon is the perimeter of the circle and the zenith is at the center as is the observer. (b) Shown schematically and in projection, as an upward-looking fisheye would see, are the 22° and 46° halos, parhelion (S), parhelic circle (PC), circumzenithal arc (CZA), pillar (P) and the circumscribed halo (CS – also known as the upper and lower tangent arcs UTA, LTA). Also shown are the sun (⊙), zenith (Z), horizon (H) and the observer (O) at the center of the celestial sphere.

5.5 The 22° halo

The 22° halo is probably the most common halo of all, occurring whole or in part many times a year every place on earth (Figure 5.5A). It appears as a thin pearly ring centered on the sun (or moon) at a distance of 22°. It has a fairly sharp reddish inner edge and a diffuse bluish outer region. Sometimes the color is hard to see and even at its best it does not rival the colors of a rainbow. The 22° halo is a few degrees wide but slightly different for each display. From the ground the complete circle of the halo appears only when the sun's altitude is greater than the angular radius of the halo and when the sky is filled with ice crystals over the extent of the halo. The 22° halo can occasionally be seen in snow as a band of sparkles (Figure 5.5B).

The 22° halo is caused by minimum deviation of light through the 60° prism in randomly oriented crystals[28,29]. This leads to its appearance as a 'hole' in the sky – a relatively dark area surrounded by a sharp, bright, reddish inner edge which gradually fades off in a blue-white haze away from the sun. The dark central hole is exactly analogous to Alexander's dark band in the rainbow (Section 4.7); a region into which light cannot be scattered by simple refraction.

Light passing at or near the angle of minimum deviation is concentrated into a narrow range of angles and is therefore bright and becomes the halo. Few crystals, however, happen to be oriented just the right way for this to occur. Most light traverses the crystal along other routes and gets scattered over such a large range of angles that it either broadens the halo or is never recognized as being part of it. Light passing through the crystal in a plane parallel to the hexagonal basal face (Figure 5.5C(a)) can either pass symmetrically (minimum deviation) or asymmetrically. Light passing through obliquely (Figure 5.5C(b)) can travel symmetrically or asymmetrically. In both cases ((a) and (b) in Figure 5.5C) the deviation is larger than 22°.

The 22° halo is actually composed of countless overlapping colored halos. The colors of halos are due to dispersion and each color has its own angle of minimum deviation (Figure 5.5D). It ranges from 21.7° in the red (656 nanometers) to 22.5° in the deep violet (400 nanometers). Because the angle of minimum deviation is smallest for longer wavelengths, these rays are deviated least of all and fall closest to the sun at the inner edge of the halo. Red rays traversing the crystals at angles that do not correspond to mini-

Fig. 5.5A The 22° halo is the most common circular halo. It is about 22° in radius. Its inner edge is relatively sharp and slightly reddish while its outer edge is bluish-white and considerably more diffuse. Depending on the crystal size, the 22° halo can be either moderately colorful or completely colorless. (Santa Monica, California)

mum deviation fall upon those rays of shorter wavelengths whose angle of minimum deviation is greater. Consequently, the colors overlap and smear the halo radially. No other color can overlap red rays at their minimum deviation and so the red inner parts of the halos have the purest colors, followed successively outwards by orange, yellow, green, blue. This is yet another reason why the reds are most visible in a halo.

Dispersion and non-minimum deviation rays are a large source of radial smearing in the halo, amounting to about 1°–2°. Owing to its 1/2° width, the sun further widens the halo uniformly in all directions[30].

Diffraction also widens the halo because crystals small enough to retain their random orientations diffract a lot of light (Figure 5.5E). The amount of broadening due to diffraction is inversely proportional to the size of the face. Any face larger than about 60 micrometers has a diffraction width of less than 1/2° and consequently has little influence on the width of the halo compared to the sun's diameter.

Like all halos, the 22° halo is polarized[31,32]. There are two reasons for this: (1) the reflection and refraction coefficients depend on polarization, and (2) birefringence, though small, is always present. The planes of polarization are not vertical or horizontal, but rather radial and tangential.

Fig. 5.5B 22° halo on the ground. Ice crystals covered the surface to return discrete spectra. The apparent circular boundary is set by the minimum deviation angle. (British Antarctic Survey)

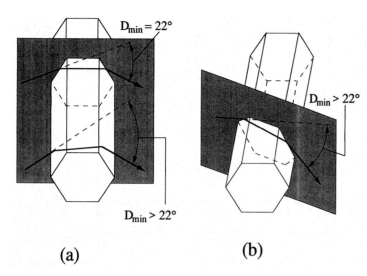

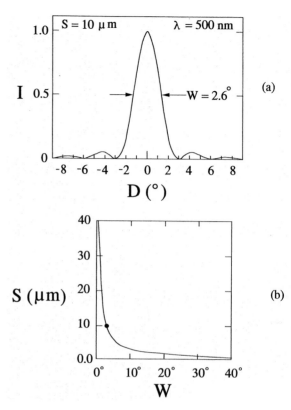

Fig. 5.5C Light can pass though the prism in three ways: (a) (1) symmetrically through the plane perpendicular to the refracting edge resulting in minimum deviation, and (2) perpendicular to the refracting edge but not at minimum deviation and (b) (3) obliquely to the perpendicular plane. The latter two paths always result in deviation larger than minimum deviation.

Fig. 5.5E (a) When collimated light falls on an aperture (in this case a crystal face) whose size is S, it is spread out into an intensity pattern whose width W is inversely proportioned to S. (b) As a result, the smaller the crystal, the wider is its diffraction width.

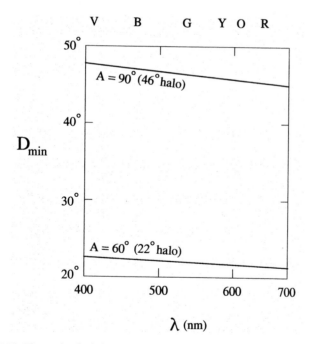

Fig. 5.5D The angle of minimum deviation varies with wavelength due to dispersion. Shown here are the angles A of minimum deviation for the 60° prism and the 90° prism.

5.6 The 46° halo

The 46° halo is similar in appearance to the 22° halo, except that it is bigger (Figure 5.6). It occurs far less frequently than the 22° halo and, because of its large size, is seldom if ever observed as a complete circle[28,29]. The 46° halo is formed by minimum deviation through the 90° prism.

Since the 22° and 46° halos can be formed in the same crystal, why is the 46° halo so rare? First, a smaller percentage of the light falling on a crystal emerges through the 90° prism. Second, dispersion spreads the 46° halo over a wider stretch of the sky making it fainter.

But there are some more subtle reasons for the 46° halo's rarity and they involve the width-to-height ratio of the crystal[33,34]. If it is long and thin like a pencil, only the 22° halo is present because so little area on the ends of the crystal is present to form the 46° halo. The crystal may also be terminated in such a way that the 90° prism is absent altogether. If the crystals are plate-like, then both halos will be present.

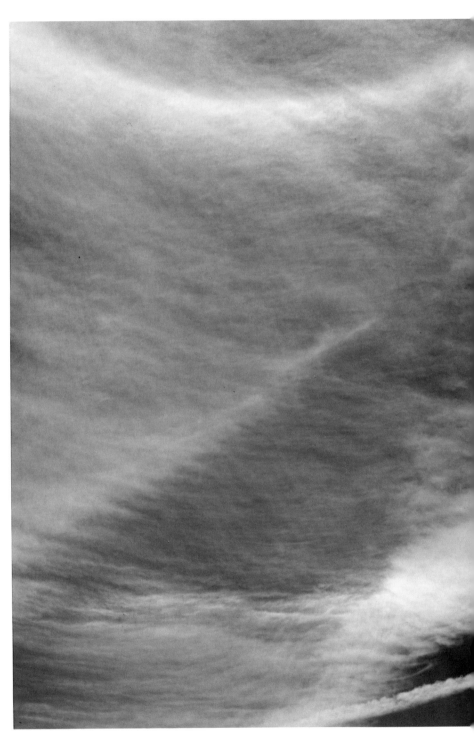

Fig. 5.6 The 46° halo (lower) and the 22° halo (upper). (Los Angeles, California)

5.7 Circular halos of unusual radii

Circular, sun-centered halos of unusual radii have been reported with virtually every radius from 4° to 40° and many beyond that[35-42] (Figure 5.7A). Those most often reported are the halos of Van Buijsen 8.3°, Rankin (Hall) 17.4°, Burney 19.0°, Dutheil 24.0°, Feuillee 32.3°, and Scheiner 28°.

Most rare halos can be explained by analyzing the pyramidally-terminated hexagonal prism, a well-known but somewhat rare ice crystal (Table 5.28C). Its most interesting property is the presence of many prisms whose refracting faces are neither 60° nor 90° apart. Pyramid crystals (often called 'bullets') can also produce halos that are nearly identical to the common 22° halo. Five of the six rare circular halos can be explained very well by this crystal.

They arise from minimum deviation through a particular prism (Figure 5.7B). The largest prism that can produce a circular halo is one whose prism angle is about 99°, although its 80° radius halo would be vanishingly faint.

Scheiner's halo of 28° has eluded explanation based on simple hexagonal pyramids[43,44]. It has been suggested that it is formed in cubic ice[45-47], a rare isometric crystal in the form of regular octahedrons. The prism angle between adjacent faces of an octahedron is 70.53°. The minimum deviation angle for such a prism is 27.46°. Such crystals are not expected to grow under normal atmospheric conditions, though they have been grown in the laboratory.

Fig. 5.7A This rare display shows the 9°,18°, 22° and 24° radii circular halos. On the original photograph, a 20° halo can also be seen. (Photo by Paul J. Neiman, Boulder, Colorado)

Hevel's halo is the largest circular halo possible[48]. With a reported radius of 90°, it follows a great circle and encircles the observer, dividing the celestial sphere into two equal halves. Hevel's halo could not be formed by simple minimum deviation since the upper limit is about 80°, and even this halo would be too faint to be detectable[18]. To form Hevel's halo, a compound mechanism requiring internal reflection is probably necessary. Without any modern photographic observations, however, Hevel's halo is likely to remain a mystery.

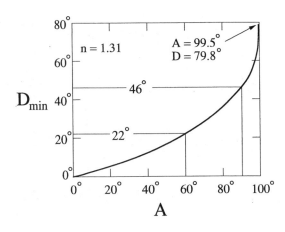

Fig. 5.7B Angle of minimum deviation D_{min} for ice ($n = 1.31$) as a function of prism angle A. The largest possible halo has a radius of about 79.8° but it would be too faint to be visible.

5.8 Parhelia (sundogs)

Parhelia are bright, colorful spots in cirrus clouds (Figure 5.8A). They are found at the same altitude as the sun and never less than 22° from it. They are usually a few degrees across and may be extended either vertically or horizontally. Parhelia usually come in pairs, one on either side of the sun, and are very common (Figure 5.8B). They are reddish on the side facing the sun and often have bluish-white tails stretching horizontally away from them. Parhelia are most prominent when the sun is low. At higher solar altitudes they grow fainter and also move away from the sun until at an altitude of about 61° they vanish altogether. When formed in moonlight they are known as paraselenae. Parhelia are sometimes seen in blowing snow.

A parhelion is formed by light that undergoes minimum deviation through the 60° prism of horizontally oriented plates[49,50] (Figure 5.8C, 5.8D). When the sun is low and the plates perfectly aligned, the parhelion is like a slice out of what would be the 22° halo if the same crystals were randomly oriented. Since all the crystals are aligned horizontally for aerodynamic stability reasons (though they are still randomly oriented about their vertical axis) parhelia tend to be far brighter than the 22° halo. They are also more colorful than the 22° halo because their oriented crystals are usually larger than those producing the 22° halo.

Sometimes the crystals are not perfectly aligned and this leads to parhelia that develop vertically along the arc of the 22° halo (Figure 5.8E). The angular height of the parhelion is then approximately equal to four times the maximum tilt angle of the crystals.

Observations of the 46° parhelia are rare and perhaps questionable. If they occur at all, they are analogous to the common parhelia discussed above: minimum deviation through the vertically oriented 90° refracting edge. Such an orientation requires two degrees of orientation for columnar crystals: (1) c axis horizontal, and (2) a axis horizontal, i.e. a pair of side faces vertical (orientation {3}). Parhelia at other azimuths relative to the sun have been reported but so far no widely recognized mechanisms to explain them have been set forth[51].

Fig. 5.8A Parhelia (sundogs) are the brightest and most common halo that originates in an oriented crystal. They are found on either side of the sun at the same altitude and tend to show bright colors. Like the 22° halo, they are red on their sunward side and fainter and more bluish away from the sun. They may show extended horizontal white tails (photo). At low solar altitudes parhelia are found 22° from the sun but at higher altitudes they separate from it. (Bishop, California)

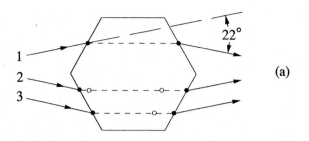

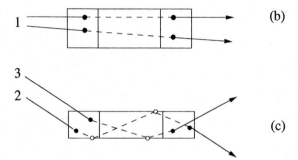

(a)

(b)

(c)

Fig. 5.8C Parhelia are due to minimum deviation through the 60° prism in ice crystals that are oriented with their *c* axis vertical. The light may be internally reflected several times before leaving the crystal.

Fig. 5.8B Parhelia often come in pairs, one on either side of the sun. Note the tails of the parhelia extending away from the sun.

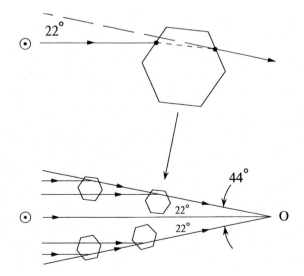

Fig. 5.8D Optics of parhelia viewed from above.

Fig. 5.8E Sometimes parhelia are extended vertically because the crystals are not perfectly oriented. This one is in blowing snow.

5.9 Paranthelia and paranthelic arcs

Paranthelia are rare parhelia-like spots a degree or so across that occur at the solar altitude and at ±120° azimuth relative to the sun (Figure 5.9A). They are colorless and may be found by themselves (very difficult to identify) or lying on the parhelic circle. Paranthelia are sometimes called parhelia of 120°.

The optical explanation for paranthelia seems straightforward: a lack of color implies either reflection or refraction through parallel faces. Their ±120° azimuth location (the exact exterior angle of an ice crystal) suggests an even number of reflections from adjacent or alternate interior side faces. That paranthelia are always located at the solar altitude means that the crystal faces responsible for them are vertical. Thus we conclude that paranthelia are formed in crystals whose c axis is vertical {1}, see Figure 5.4C.

There are at least two ways to make a paranthelion and they both probably play a role in its formation, one being more important when the sun is high, the other when the sun is low (Figure 5.9B). At high solar altitudes, light enters the upper basal faces, reflects twice off adjacent or alternate vertical side faces, then leaves the crystal via the lower horizontal basal face. At low solar altitudes, light enters the crystal through a side face, reflects

twice, then leaves through a side face. The terms 'high' and 'low' solar altitudes are relative terms which depend largely on the aspect ratio of the crystals.

Paranthelic arcs are short, nearly vertical arcs that pass through the paranthelia. These rare arcs are thought to be due to the same crystal orientations that produce paranthelia except the crystals are not perfectly aligned. This causes the paranthelia to be smeared vertically, resulting in the arcs.

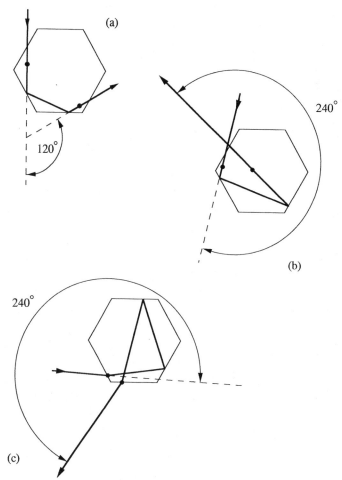

Fig. **5.9A** Paranthelia are found at the solar altitude but at azimuths of ±120°. They are colorless spots and may be found either on the parhelic circle or with oblique arcs passing through them. The parhelic circle is a colourless horizontal arc that may encircle the entire sky. It also forms at the same altitude as the sun.

Fig. **5.9B** The paranthelia are formed in at least three ways. Light enters the top surface and reflects from either adjacent (a) or alternate (b) side faces before leaving the crystal through the bottom surface. It can also enter and exit through adjacent side faces (c). In each case the deviation is either 120° or 240°.

5.10 Circumzenithal arc and circumhorizontal arc

The circumzenithal and circumhorizontal arcs (CZA and CHA respectively) are perhaps the loveliest halos of all. Their vivid colors are surpassed only by the exceptional parhelia. For sheer size and brilliance, only the rainbow comes close to matching their splendor.

The CZA (also called the Bravais' arc) is found almost directly overhead in the form of a quarter of a circle centered on the zenith and facing the sun (Figure 5.10A). The colors run from red to blue with increasing altitude. It is most often reported when the sun is at an altitude of approximately 22°. It is then about 22° from the zenith and about 3° across. It is regularly associated with other halos, especially the 46° halo, to which it often appears tangent.

The CHA is a brilliantly colored band of light parallel to the horizon (red above, blue below) which is always found on the same side of the sky as the sun (Figure 5.10B). Reports of the CHA are rare because it is found low in the sky where it may be obscured by the landscape.

Both halos are formed by crystals oriented with their c axes vertical {1} and with their 90° refracting edges horizontal[52] (Figure 5.10C). For the CZAs, light enters the upper horizontal basal face and exits through one of the vertical side faces. The reverse is true for the CHAs: entry is through a vertical side face and exit through the lower horizontal basal face. Even though neither halo is a minimum deviation phenomena, they both are brightest when light traverses the crystal at minimum deviation. For the CZA this happens when the solar altitude is 22.13°, at which point the halo is exactly tangent to the 46° halo. The same idea explains why the CHA should be brightest at a solar altitude of 67.87°, where again the halo is tangent to the 46° halo. For solar altitudes between 5° and 30° and for perfectly oriented crystals, the width of the CZA varies between 2° and 6°, and its azimuthal extent between 58° and 78°.

The CZA and CHA are closely related in their form and occurrence (both being parallel to the horizon). The sun must be lower than 32.2° altitude for a CZA. The CHA appears only when the sun is higher than 57.8° (= 90°−32.2°), and therefore the two halos can never be seen at the same time. The CZA can, in principle, be observed everywhere on earth at one time of the year or another. The CHA, however, is confined to latitudes of less than about 56° because at greater latitudes the sun can never reach the required altitude of 57.8°.

Calculations show that the CZA changes with solar altitude[19]. When the sun is on the horizon, the CZA reaches its largest possible distance of about 32° from the zenith, but is too faint to be seen. As the sun's altitude increases the halo brightens, widens, and moves closer to the zenith until maximum visibility occurs at a solar altitude of 22°. At higher altitudes the CZA widens, fades and ultimately disappears as it collapses into the zenith when the sun is at 32.2°.

The CZA and CHA are associated with the 46° halo because they are all formed in the horizontally-oriented 90° refracting edge. Precisely aligned crystals generate arcs with brilliant colors while slightly wobbly crystals produce wider, somewhat more pastel arcs.

Kern's arc[53] is a rare attendant to the CZA. It appears opposite the CZA as though completing the circle centered on the zenith. Kern's arc is thought to form when part of the light striking the inner vertical crystal face is internally reflected until it emerges through an opposite vertical face. This arc should be much fainter than the CZA because of the one or possibly three additional reflections, none of which is total.

LOOK UP FOR HALOS

Certain exceptional halos may be found more-or-less directly overhead. The best one is the circumzenithal arc which can easily escape our attention unless we make a point of the somewhat awkward movement of glancing up when we go outdoors. Any sun-centered phenomenon such as the 22° halo will also be high in the sky when the sun is high. Most often we scan the sky and it does not occur to us to inspect the zenith. Try it.

Fig. 5.10A The circumzenithal arc is found high overhead when the sun is less than about 32° above the horizon. It is a brightly colored arc centered on the zenith and facing the sun and is brightest when the sun is about 22° above the horizon. The blue side of the arc is nearest the zenith and the red side nearest the sun. (Photo by Paul Neiman)

Fig. 5.10B Circumhorizontal arc. (Photo by Barry Jones)

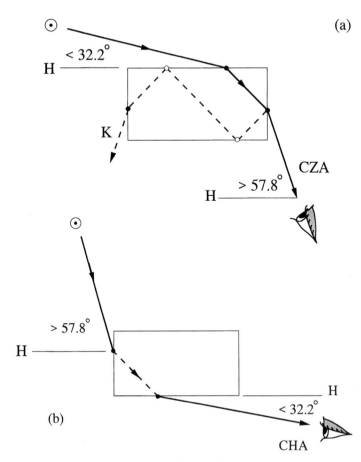

(a)

(b)

Fig. 5.10C The circumzenithal arc and circumhorizontal arc are complementary halos. The CZA is formed by deviation of light passing through the 90° prism of plates whose *c* axes are vertical (a). Light enters the upper basal face and exits through a side face. Because the crystals are randomly oriented around the *c* axes, the halo is spread out along a circular arc. The CHA is formed in exactly the same way, except light enters a side face and leaves through the lower basal face (b).

5.11 Pillars

Shafts of light that extend vertically from the sun are called pillars (Figure 5.11A). They are most often seen above the sun when it is low, especially when it is within 1° or 2° of the horizon. In cold regions where ice fogs and snow occurs, pillars can often be seen standing over street lights. Their color mimics that of sunlight (or the source).

Pillars are formed by grazing reflections from the lower horizontal faces of plates oriented with their *c* axes vertical[54-6] (Figure 5.11B). Light that is refracted upwards through the lower face and then reflects back down from the upper face before leaving the crystal through the lower face also produces pillars. Plates roughly 1 millimeter across can attain nearly perfect orientation which results in thin, bright pillars. Any deviations from strict alignment broadens the pillar. Nearby pillars seen over street lights can vanish and reform with remarkable speed as gusts of wind momentarily destroy crystal alignment.

Pillars originating at great distances require the crystals to be slightly misaligned because with precise horizontal orientation the light is either directed upwards (and hence is inaccessible to the observer) or passes through the crystal undeviated (appears to come from the sun). Only when the crystals are tilted slightly can sunlight reflect from the lower surfaces. The maximum tip of the crystals is about half the angular length of the pillars; a 10° high pillar is produced in crystals that are tipped by up to about 5°. If this pillar was seen from the air, it would appear both above and below the sun and span four times the maximum inclination or 20°. Nearby pillars do not require perfectly aligned crystals. Street light pillars are almost totally polarized at the Brewster angle[57]. Natural pillars are almost certainly polarized in the same way but they are too faint at large altitudes to have been measured. From the air, pillars can be seen below the sun.

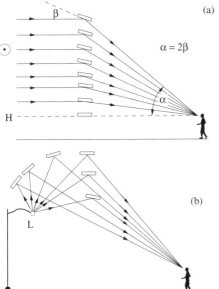

Fig. 5.11A Pillars are narrow vertical shafts of light extending a few degrees from the sun, usually above it. They are normally seen when the sun is low on the horizon and mimic the sun's color. (Pontatoc Canyon, Arizona)

Fig. 5.11B Pillars are due to the reflection of sunlight from the undersides of oriented plates that are slightly tipped about the vertical direction (a). The total height above the sun reached by the pillar (α) is equal to about twice the average tip angle β. Light that passes through the crystal by reflecting from the upper surface also contributes to the pillar. Although a slight tilt of the crystals is required to see a sun pillar, even perfectly aligned plates can cause a pillar when the light source is nearby as the two crystals at the top and right show (b).

5.12 Parhelic circle

The parhelic circle is a horizontal band of white light a degree or so high that lies at the solar altitude and extends for many degrees around the sky[58,59] (Figure 5.9A). If cirrus covers the sky uniformly, the halo may completely encircle the observer. More often it appears only in part. When it forms, it passes through the sun, the parhelia and paranthelia if present, and the anthelic point. Though not a common halo, the parhelic circle may be overlooked due to variable cloud spacing and thickness and because the thin colorless halo may masquerade as a bit of horizontally-extended cirrus.

The parhelic circle is formed when sunlight reflects from the front surface of any crystal face that lies in the vertical plane (Figure 5.12). Thus it is due to oriented crystals. The parhelic circle can be produced by all three types of oriented crystal because each has a vertical face. With no way of orienting the horizontal axes in azimuth, they are randomly distributed and reflect light in every azimuthal direction. Such crystals could also produce a parhelic circle by transmitting light and reflecting it off a rear surface. Like parhelia, light from the parhelic circle can be piped along the inside of the crystal.

5.13 Tangent arcs of the 22° halo (circumscribed halo of 22°)

The tangent arcs of 22° appear as two separate arcs when the sun is low and as a circumscribed halo when the sun is high (Figures 5.13A, 5.13B). They are tangent to the 22° halo at its highest and lowest points. The upper and lower tangent arcs may appear as brightenings of the 22° halo. As the solar altitude increases, the two arcs 'reach' towards one another and join when the sun is around 30° to form a droopy circumscribed halo which remains tangent to the 22° halo (Figure 5.13C). As the sun rises in the sky, the halo grows smaller and more symmetric, ultimately collapsing into the 22° halo at a solar altitude of 70°.

The crystals responsible for the tangent arcs are columns oriented with their c axes horizontal but with random orientations of their a axes[60]. Such crystals are often referred to as 'spinning' (Figure 5.13C(c)). Light enters one side face and exits through an alternate side face. Since these are the same 60° prisms that cause the 22° halo, it is easy to see why the circumscribed halo is tangent to the 22° halo at its horizontal points: the spin axis is horizontal. At the tangent points the 22° halo and tangent arcs are identical. Rays traversing the crystal away from the principal plane encounter a prism whose apex angle is greater than 60° and thus their minimum deviation angle is also greater than 22°. This explains why the circumscribed halo is always further from the sun than the 22° halo.

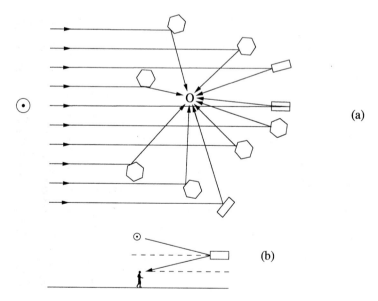

(a)

(b)

Fig. 5.12 The parhelic circle is formed by external reflection from crystal faces lying in the vertical plane. Under some circumstances the light may also pass through the crystal without undergoing dispersion. In this view from above, the observer O may see himself surrounded by the parhelic circle.

Fig. 5.13A The circumscribed halo is also called the upper and lower tangent arcs of the 22° halo when the sun is low. For a high sun it appears as an elliptical halo that is outside the 22° halo except at its upper and lower points where it is tangent. At lower altitudes the lobes of the ellipse begin to droop and fade, ultimately appearing only directly above and below the sun and tangent to the 22° halo. Also visible is the parhelic circle and a parhelion. (Photo E. Richard Norton)

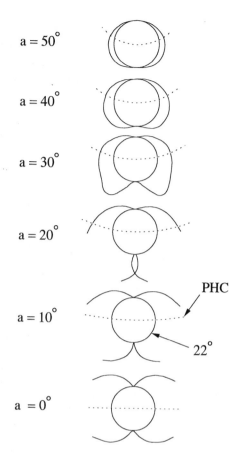

$a = 50°$

$a = 40°$

$a = 30°$

$a = 20°$

PHC

$a = 10°$

$22°$

$a = 0°$

Fig. 5.13B The circumscribed halo changes shape as a function of solar altitude. PHC is the parhelic circle.

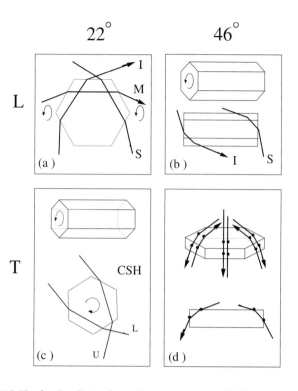

$22°$ $46°$

L   (a)   (b)

I   M   S   I   S

T   CSH   U   (c)   (d)   L

Fig. 5.13C The family of halos formed by spinning crystals. They are known as lateral (L) and tangent (T) arcs of the 22° and 46° halos because they touch them. (a) There are three Lowitz arcs. The supralateral (S) and infralateral (I) arcs are the only two to be seen. The third mesolateral arc (M) has only been predicted by computer simulations. (b) There are two lateral arcs of the 46° halo, the infralateral (I) and supralateral (S). (c) The circumscribed halo (CSH) is caused by refraction through the 60° prism in spinning columns whose c axes are horizontal. The azimuths of the columns are randomly oriented. The circumscribed halo is also formed by columns that are randomly oriented about the horizontal c axis (optically equivalent to spinning crystals). (d) Tangent arcs of the 46° halo have never been seen although six are predicted.

5.14 Subsuns and Bottlinger's rings

Subsuns are vertically-elongated ellipses seen from aircraft when flying above cirrus clouds[61] (Figure 5.14A). They are found directly below the sun at the same angle below the horizon as the sun is above it. The subsun is often exceedingly bright and occasionally reproduces the sun with enough fidelity to be perfectly round with sharp edges.

The subsun is an image of the sun reflected from the upper side of any crystal with a horizontal face. The crystal most often identified with the subsun is a horizontally oriented plate. When the crystals are slightly out of perfect alignment, the image of the sun becomes slightly elliptical. It is elongated in the vertical direction in precisely the same way that the glitter is elongated. In principle the subsun should also originate by transmission into the crystal, reflection from the lower face and then transmission from the upper face. A bright subsun could produce its own halos but such halos would probably be too faint to be noticeable.

Bottlinger's rings are sometimes observed around the subsun when it is slightly elliptical[61] (Figure 5.14B). They appear as a single elliptical ring surrounding the subsun and they are rare. The origin of Bottlinger's rings is uncertain but they may be due to either oscillations or gyrations of the same crystals that cause the subsun: horizontally oriented plates.

Off the north coast of Japan the extremely brilliant lights of a night fishing fleet have been seen mirrored in the sky by these same horizontally oriented crystals[62]. The display is like multiple subsuns except these are viewed from below.

Fig. 5.14A The subsun is a bright colorless spot seen below the horizon on the solar vertical circle when flying over cirrus clouds. It is found at the same distance below the horizon that the sun is above it. The subsun may be slightly elongated in the vertical plane. It is simply the reflection of the sun from horizontally oriented faces, usually plates. The ellipticity of the subsun is due to slight departures of the reflecting surfaces from horizontal. This vertical extension of the image of the sun occurs in exactly the same way that glitter becomes vertically elongated. (Iceland)

Fig. 5.14B Bottlinger's rings are faint, colorless elliptical rings a degree or so in diameter that encircle the subsun. They are extremely rare. (Photo by Fred Mertz, Australia)

5.15 Halos below the horizon

There are two classes of halo that appear below the horizon, implying observation from an airplane: (1) those that can only be found there such as the lower tangent arc of the 22° halo when the sun is lower than 22° above the horizon, and (2) those halos easily recognized because they are obviously reflections of common halos normally seen above the horizon. At one time the latter halos were seldom seen because of the unlikely chance of an observer being above an ice cloud. With the arrival of jet air travel and thousands of people flying above cirrus clouds daily, subhalos have become a common sight.

Any crystal that has a horizontal face can produce a halo below the horizon by reflecting light upwards from either its upper exterior face or its lower internal face. Those halos which can occur below the horizon are noted in Table 5.28B. The known subhalos that appear at their usual azimuths but below the horizon are the subsun, subpillar, subparhelic circle, and the subparhelion.

Subanthelic arcs always occur below the horizon[19] and the rare occasions of their sightings have always been from aircraft. They look very similar to the anthelic arcs except the arcs cross at the antisolar point. They are formed in horizontally oriented columns.

RARE AND UNUSUAL HALOS

5.16 Parry arcs

Parry arcs are two short arcs that are found immediately above and below the 22° halo[63]. The upper Parry arc is the most frequently reported of the two, appearing as a short arc that is concave downward just above the 22° halo (Figure 5.16). In conjunction with the circumscribed halo (upper tangent arc of 22°), the upper arc forms a football-shaped 'eye' at some solar altitudes. The lower arc is tangent to the lowest part of the 22° halo. Both change shape with solar altitude. Parry arcs were named for their discoverer, Admiral Parry, during his search for the northwest passage.

Parry arcs arise from ice columns oriented with their c axes horizontal and a pair of opposite side faces horizontal. Computer analyses of Parry arcs reveal that there are actually four arcs, two above and two below[64]. Because the pairs of arcs overlap, it is difficult to distinguish them from each other.

Fig. 5.16 Parry arc. The most commonly observed of the four possible Parry arcs is the upper concave Parry arc. With the circumscribed halo below, it forms the shape of a football. (Photo by Jack Harvey at the South Pole)

5.17 Heliac arcs

The same crystals responsible for Parry arcs are expected to produce another halo called the heliac arc. It has not been well observed and possibly not at all. The heliac arc takes on a complex shape that stretches over the entire sky[19] and changes markedly with solar altitude. Its presumed origin is a single external reflection from columns oriented with their c axes horizontal but randomly oriented in azimuth. Such a halo would be colorless.

5.18 Anthelion

The anthelion or 'counter-sun' is a rare, colorless spot a degree or so across located at the anthelic point. It is usually seen at the intersection of the anthelic arcs or in rare instances sitting alone or on the parhelic circle. No good photographs are known.

Many crystal shapes and orientations can cause this rare halo[65-7]. Columns with both c and a axes horizontal {3} would easily create the anthelion (Figure 5.18). Although this mechanism successfully accounts for the lack of anthelion sightings when the sun is higher than 46°, the orientation is unlikely. Computer simulations of the anthelion suggest that it may not exist as a separate halo but is in reality simply the accumulation of light from the intersecting anthelic arcs. Calculations by Greenler[19] also suggest that it may be a form of the anthelic pillar.

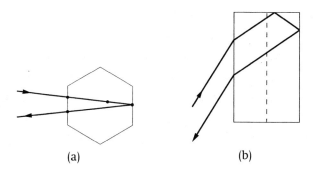

(a)   (b)

Fig. 5.18 The simplest optical path that light can take to form an anthelion is by internal reflection from a 90° prism whose refracting edge is vertical. Light enters and leaves the crystal through the same vertical face. This requires the unlikely {3} orientation: (a) side view, (b) top view.

5.19 Anthelic pillar

The anthelic pillar is a colorless vertical shaft of light a few degrees tall that passes through the anthelic point. So few accurate observations have been made that little more can be said about its properties.

A halo that resembles the anthelic pillar has been found using computer simulations[19]. The crystal is a horizontally oriented column with a pair of vertical side faces {3}. There are three possible optical schemes for the anthelic pillar, each being most important at certain solar altitudes. The process that is common to all three regimes is two successive internal reflections from an end and a side face. Since these faces are 90° apart, the deviation is 180°. It is possible that the anthelic pillar is just an anthelion formed in crystals that are not perfectly oriented.

5.20 Tangent arcs of the 46° halo

Tangent arcs of the 46° halo (also called contact arcs) have not been directly identified in nature. Their existence and shapes are inferred from computer simulations[68]. These numerical experiments show the shapes and locations of the arcs, as well as their properties which change with solar altitude. There are six arcs distributed around the outside of the 46° halo and tangent to it at various locations. The fact that they have not been unambiguously observed may explain some of the mysteries of the 46° halo's properties. The 46° may never appear uniform in brightness because the faint tangent arcs are contributing their own light, which is brighter at some locations than others.

The crystals responsible for the arcs are plates spinning about a horizontally oriented a' axis and each comes from a ray which passes through adjacent side and end faces (Figure 5.13C(d)). Each crystal has three such pairs of faces, two adjacent to the spin axis but at different angles to it, and a pair that is parallel to the spin axis. Light can pass through each pair of three faces two ways (entering through a side face and leaving through an end face – and vice versa) resulting in six arcs.

5.21 Lateral arcs of the 22° halo (Lowitz arcs)

There are, in principle, three lateral arcs of the 22° halo, but only two have been observed. They all pass through the parhelia which gives an important clue as to their origin. The supralateral arc is tangent to the 22° halo above the sun and the infralateral arc is tangent below. An as yet undetected (but theoretically possible) mesolateral arc passes approximately vertical through the parhelion. The 22° lateral arcs are often called Lowitz arcs after Tobias Lowitz who first observed them during a spectacular halo display in 1790 in St Petersburg, Russia[69].

Lowitz arcs are rare halos formed in plates spinning about their horizontal *a* axis[70] (Figure 5.13C(a)). Light enters a side face and leaves through an alternate side face, thereby traversing the crystal through the 60° prism. Three arcs occur because there are three different ways light can pass through the crystal. Their intersection at the parhelion takes place when the crystals are horizontal, and thus momentarily identical to parhelia-producing plates {1}. Their color distribution is similar to the parhelion's.

5.22 Lateral arcs of the 46° halo

Lateral arcs of the 46° halo are rare, though a few photographs do exist. They are formed in the same crystals that produce the circumscribed halo, although the optical mechanism involves the 90° prism rather than the 60° prism[71] (Figure 5.13C(b)). The upper arc is called the supralateral arc and the lower one the infralateral arc.

The lateral arcs of the 46° halo are closely related to the CZA and CHA. Since both sets of arcs require the 90° prism with its apex horizontal, they have similar visibility limits. The infralateral arc should be visible at any solar altitude but the supralateral arc can only be seen when the sun is higher than about 32°.

5.23 Anthelic arcs

A number of rare, thin arcs called anthelic arcs have been reported and photographed that cross the sky and point towards the sun at one end and the anthelic point at the other[72-4] (Figure 5.23). Most show some color.

There seem to be a number of different optical paths that can produce anthelic arcs and all involve a horizontally oriented column[75,76]. At any one time they all may contribute to the formation of arcs so it is not possible to uniquely identify the optical paths. In other words, anthelic arcs are actually composite halos that are formed by several mechanisms operating simultaneously that produce very similar, and to the eye indistinguishable, halos. The paths suggested for the different arcs are listed in Table 5.28E. They are responsible for Wegener's arcs, Tricker's arcs, and the so-called diffuse A and diffuse B arcs. Like the CZA, none are minimum deviation phenomena but may well be brightest at the minimum deviation angle.

Fig. 5.23 Anthelic arcs and the parhelic circle intersecting at the anthelic point. They have been reported in a number of configurations and may extend across the sky and well into the hemisphere containing the sun. (Photo by O. Richard Norton)

5.24 Multiple halo displays

When more than one halo is seen it is called a halo display. Multiple halos mean that different types of crystals are present (Figure 5.24). Since most halos come from a single form of crystal, a complex halo display is convincing evidence that a wide variety of crystals can occur simultaneously along the line-of-sight.

A particularly complete record of halo displays was given by Liljequist[71] who reported a tremendous variety of halo data including drawings, weather records, *in situ* crystal samples, frequency statistics and some of the best shape measurements of Parry arcs, Hall's halo, paranthelia and the upper tangent arc. In his splendid book *Scrambles Amongst the Alps*, Edward Whymper[77] describes a cross seen in the clouds. This was almost certainly the superposition of a pillar and a parhelic circle.

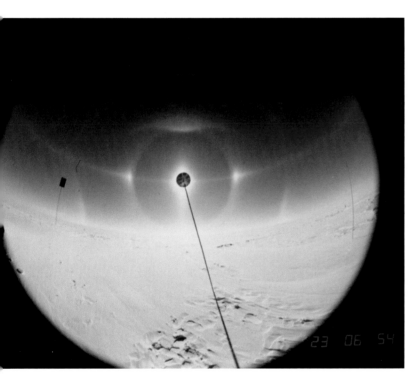

Fig. 5.24 A halo display showing the 22° halo, the 46° halo, the upper tangent arc, Parry arc, CZA and the parhelic circle. What can we deduce about the crystals in the clouds? (Photo by G. Konnen)

5.25 Multiple scattering effects

All of the halos described so far have been explained on the basis of a single interaction of a light ray with a crystal. In reality the light scattered by the crystal will probably strike another crystal and be scattered again. Further scattering may also occur. Because the scattered light will no longer come from the sun's direction and will also appear to come from the halo produced by the first scattering, a second scattering will result in an extremely complex halo display that may be too faint to perceive.

There are, however, certain multiple scattering halos that may be detectable under the right circumstances[78,79]. The most likely is the parhelion produced by the parhelion of the 22° halo, which appears as a faint parhelion 44° from the sun. This halo has been reported in the famous halo display at Saskatoon[80,81] in 1970. Other possible multiple scattering halos are the 66° parhelia, the anthelion and a colorless CZA.

5.26 Diffraction halos

Many of the crystals associated with halos are so small that a significant fraction of the light scattered from them is due to diffraction. The usual effect is to broaden the halo slightly and raise the level of background light. When the crystals are oriented, however, they may produce a cross-like halo centered on the sun[82].

5.27 Elliptical halos

Early this century[83-6] a handful of elliptical halos were reported but it wasn't until 1988 that they were verified photographically[84-6]. All elliptical halos reported so far are taller than they are wide and are centered on the sun or moon. And all are very rare. The elliptical halo most frequently reported, called Schlesinger's halo, is about 7° by 4°. A larger halo, 10° by 5°, has become known as Hissink's halo. A third as yet unnamed halo is roughly 40° by 30°.

Some of the halos show colors, other do not. The absence of color may be a consequence of moonlight observations. Owing to the sparse number of sightings, we do not know if elliptical halos come in certain well-defined sizes or whether there is a continuum of sizes. We also have little idea how they are formed. It is even possible that ice crystals are not involved[61, 87-9].

5.28 Halo catalog

Halos can be classified according to the prism angle that produces them (Table 5.28A). Owing to the large number of halos, it is also useful to classify them according to orientation and spinning configurations (Table 5.28B). Tables 5.28C, 5.28D, and 5.28E explicitly list the path taken by the ray which produces the halo.

Table 5.28A. *Two-family classification of major halos.*

Optics	60° Prism	90° Prism
minimum deviation through randomly oriented crystals	22° halo	46° halo
minimum deviation through crystals with refracting edges vertical	22° parhelion	[46° parhelion]
deviation through crystals with refracting edges horizontal	upper Parry arc / lower Parry arc	circumzenithal / circumhorizontal
double internal reflection from vertical faces of oriented crystals	paranthelion	[anthelion]
deviation through crystals spinning about a horizontal *c* axis	circumscribed halo (upper and lower tangent arcs)	infralateral arcs supralateral arcs
deviation through crystals spinning about a horizontal *a′* axis	infralateral arc mesolateral arc supralateral arc	[46° tangent arcs]

[ ] rarely and perhaps never observed

Table 5.28B. *Halo–crystal cross reference.*

Orientation	Halos	
random	22° halo 46° halo all circular, sun-centered halos	
c axis vertical *a* axis horizontal {1}	22° parhelion circumzenithal arc circumhorizontal arc parhelic circle paranthelia pillar	
c axis horizontal	circumzenithal arc	possible subhalos
a axis vertical {2}	circumhorizontal arc Hasting's anthelic arc parhelic circle Parry arcs pillar heliac arcs	
c axis horizontal *a* axis horizontal {3}	anthelion 46° parhelion parhelic circle	
spinning about horizontal *a′* axis spinning about horizontal *c* axis	22° lateral arc 46° tangent arc 22° tangent arcs 46° lateral arcs Wegener's anthelic arcs Tricker's anthelic arcs diffuse A and B anthelic arcs pillar parhelic circle anthelion anthelic pillar	

{1} optically equivalent to spinning about a vertical *c* axis
{2} optically equivalent to spinning about a vertical *a* axis
{3} optically equivalent to spinning about a vertical *a′* axis

Table 5.28C. *Ray paths for randomly oriented crystals.*

Crystal	Halo or effect	Ray paths
right hexagonal prism		
	glint	1,2,3, etc.
	sparkle	13,37, etc.
	22° halo	37,48, etc.
	46° halo	13,17, etc.
pyramidally terminated right hexagonal prism		
	Van Buijsen	6F3
	Rankin (Hall)	F4R6
	Burney	F3F6
	Dutheil	6F4
	Feuillee	F4F6

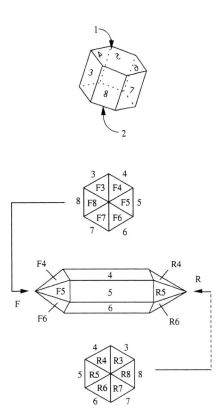

Table 5.28D. *Halo–ray paths for plates.*

Orientation	Halo or effect	Ray paths
c axis vertical		
{1}		
	22° parhelia	37,48, etc.
		3217, etc.
	circumzenithal arc	13,18, etc.
	circumhorizontal	32,82, etc.
	parhelic circle	3,8,7, etc.
		3216
wobbling	pillar	2,212, etc.
	paranthelia	1382
		1372
		3758
wobbling	paranthelic arcs	1382
		1372
		3758
	Kern arc	13216
wobbling	subsun	1
wobbling	Bottlinger ring	1
	subparhelia	327, etc.
spinning about *a′* axis		
S{1}		
	46° tangent arcs	13,17, etc.
		31,71, etc.
		18,81, etc.
	Lowitz arcs 22°	
	infralateral	68,75
	supralateral	57,86
	mesolateral	46,37

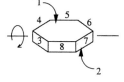

Table 5.28E. *Halo–ray paths for columns.*

Orientation	Halo	Ray paths
c axis horizontal a axis vertical {2}		
	circumzenithal arc	31,32
	circumhorizontal arc	16,26
	circumscribed halo 22°	
	upper tangent	84
	lower tangent	75
	Parry arcs	
	upper suncave	35
	upper sunvex	84
	lower suncave	86
	lower sunvex	75
	parhelic circle	1,2
		1632
	heliac arcs	7,5
wobbling	pillar	6,636
wobbling	subsun	3,363
c axis horizontal a′ axis vertical {3}		
	parhelic circle	1,2,5,8
	anthelion	8158
	anthelic pillar	
	low sun	8518
	med sun	3517
	high sun	1521
	46° parhelia	81,15, etc.
c axis horizontal spinning about c axis S{2}		
	lateral arcs 46°	
	infralateral	16
	supralateral	32
	anthelic arcs	
	Wegener	
	Tricker	15321
		81463
	diffuse A	824568
	diffuse B	8146
	parhelic circle	1,2

THE HIGHEST CLOUDS

5.29 Noctilucent clouds ('night-shining' clouds)

Noctilucent clouds are rare, bluish or whitish and are visible against the night sky long after twilight. They are mainly seen at high latitudes during summer. They look like cirrostratus clouds but instead of brightening as dawn approaches, they disappear[90].

We know that noctilucent clouds are very high because they remain illuminated by the sun long after the sunset on the ground. Triangulation tells us their elevation is about 85 kilometers near the mesopause. The air here is cold (−100 °C) and thin (1/1000th of 1% of sea-level pressure). Rocket measurements show that noctilucent cloud particles are small (~0.3 micrometers) and composed of ice that has condensed on extraterrestrial (meteoritic) dust particles. The source of the water vapor is surely tropospheric but how it reaches elevations of 80 kilometers is still poorly understood.

Fig. 5.29 Noctilucent clouds over Calgary, Canada. (Photo by A. Clark)

5.30 Nacreous clouds

Nacreous clouds are very similar to noctilucent clouds except that they form in the low stratosphere at heights of about 20–30 kilometers. Nacreous clouds can occasionally be seen during the day and resemble thin cirrus clouds often with strong irisation (Figure 5.30). As the sun sets, nacreous clouds remain bright while tropospheric cirrus grows dark and is seen silhouetted against the twilight. Little is known about nacreous clouds and how they form.

Fig. 5.30 Nacreous clouds, teaming with iridescence, over Ross Island, Antarctica. (Photo by B. McKibben)

References

1 Wiscombe, W.J. and Warren, S.G., 1980, 'A model for the spectral albedo of snow I: Pure snow', *Journal of Atmospheric Science*, **37**, 2712–2733.

2 Bentley, W.A. and Humphreys, W.J., 1931, *Snow Crystals*, McGraw-Hill Book Company, New York (Bentley, W.A. and Humphreys, W.J. 1962, *Snow Crystals*, Dover, New York).

3 LaChapelle, 1969, *Field Guide to Snow Crystals*, University Washington Press, Seattle.

4 Runnels, L.K., 1966, 'Ice', *Scientific American*, **215**, No.6, 118–126.

5 Kamb, B., 1973, 'Crystallography of ice' in *Physics and Chemistry of Ice*, E. Whalley, S. Jones and L. Gold, Eds., Royal Society of Canada, Ottowa, 28–41.

6 Magono, C. and Lee, W.C., 1966, 'Meteorological classification of snow crystals', *Journal of the Faculty of Science*, Hokkaido Univ. Ser. VII. (Geophysics), II, no.4.

7 Nakaya, U., 1954, *Snow Crystals: Natural and Artificial*, Harvard University, Cambridge.

8 Colbeck, S.C., 1982, 'Overview of seasonal snow meta-morphism', *Reviews of Geophysics and Space Science*, **20**, 45.

9 Warren, S.G., 1982, 'Optical properties of snow', *Reviews of Geophysics and Space Science*, **20**, 67.

10 Warren, S.G., 1984, 'Optical constants of ice from the ultraviolet to the microwave', *Applied Optics*, **23**, No. 8 April, 1206–1225.

11 Bohren, C.F. and Barkstrom, B.R., 1974, 'Theory of the optical properties of snow', *Journal of Geophysics Research*, **79**, 4527–4535.

12 Bohren, C.F., 1983, 'Colors of snow, frozen waterfalls and icebergs', *Journal of the Optical Society of America*, **73**, 1646–1652.

13 Lee, R., 1990, 'Green icebergs and remote sensing', *Journal of Optical Society of America A*, **7**, 1862–1874.

14 Pollock, R., 1970, 'What colors the mountain snow?', *Sierra Club Bull.* **55**, 18 (#4).

15 Pernter, J.M. and Exner, F.M., 1922, *Meteorologische Optik*, Wilhelm Baumuller, Vienna and Leipzig.

16 Tricker, R.A.R., 1970, *Introduction to Meteorological Optics*, American Elsevier, New York.

17 Tricker, R.A.R., 1979, *Ice Crystal Halos*, Optical Society of America, Washington DC.

18 Lynch, D. K., 1978, 'Halos', *Scientific American*, **238**, no.4, 144–152.

19 Greenler, R.G., 1980, *Rainbows, Halos and Glories*, Cambridge University Press, Cambridge.

20 Tape, W., 1994, *Atmospheric Halos*, v. 64 of Antarctic Research Series, American Geophysical Union, Washington.

21 Hastings, C.S., 1920, 'A general theory of halos', *Monthly Weather Reviews*, **48**, 322–330.

22 Tape, W., 1979, 'Geometry of halo formation', *Journal of the Optical Society of America*, **69**, 1122–1132.

23 Tape, W., 1982, 'Folds, pleats and halos', *Am. Sci.*, **70**, 467–474.

24 White, R., 1976, 'An analytic theory of certain halo arcs', *Journal of the Optical Society of America*, **66**, 768–772.

25 Konnen, G.P., 1983, 'Polarization and intensity distributions of refraction halos', *Journal of the Optical Society of America*, **73**, 1629–1640.

26 Cho, H.R., Iribane, J.V., and Richards, W.G., 1981, 'On the orientation of ice crystals in a cumulonimbus cloud,' *Journal of Atmospheric Science*, **38**, 1111–1114.

27 Edge, R.D., 1976, 'Bernoulli and the paper dirigible', *American Journal of Physics*, **44**, 780–781.

28 Fraser, A.B., 1979, 'What size of ice crystals causes the halos?', *Journal of the Optical Society of America*, **69**, 1112–1118.

29 Weinheimer, A.J., 1986, 'What size of ice crystals causes the halos?: Comment', *Journal of the Optical Society of America* A, **3**, 376–377.

30 Lynch, D.K. and Schwartz, P., 1983, 'Intensity profile of the 22° halo', *Journal of the Optical Society of America* A, **2**, 584–589.

31 Konnen, G.P., 1977, 'Polarization of halos and double refraction', *Weather*, **32**, 467–468.

32 Konnen, G.P., 1991, 'Polarimetry of 22° halo', *Applied Optics*, **30**, No.24, 3382–3400.

33 Greenler, R.G., Mueller, J.A., Hahn, W., and Mallmann, A.J., 1980, 'The 46° halo and its arcs', *Science*, **206**, 643–649.

34 Pattloch, F. and Trankle, E., 1984, 'Monte Carlo simulation and analysis of halo phenomena', *Journal of the Optical Society of America* A, **1**, 520–526.

35 Andus, C.G., 1915, 'Solar halo of May 11, 1915 at Sand Key, Florida', *Mon. Weather Reviews*, **43**, 213–214.

36 Besson, L., 1923, 'Concerning halos of abnormal radii', *Monthly Weather Reviews*, **51**, 254–255.

37 Besson, L. and Dutheil, X., 1900, *Annales de l'Observatoire Municipe de Montsourie*, **4**, 290.

38 Bravais, A., 1847, 'Mem. sur les halos et les phenomenes optiques', *J. Ec. Roy. Polytech.*, **18**, 1.

39 Goldie, E.C.W., Meaden, G.T., and White, R., 1976, 'The concentric halo display of 14 April 1974', *Weather*, **31**, 304–312.

40 Humphreys, W.J., 1922, 'Certain unusual halos', *Monthly Weather Reviews*, **50**, 23–28.

41 Humphreys, W. J., 1929, *Physics of the Air*, McGraw-Hill, New York. Reprinted 1963, Dover, New York.

42 Neiman, P., 1989, 'The Boulder, Colorado concentric halo display of 21 July 1986', *Bulletin of the American Meteorological Society.*, **70**, No. 3, 258–264.

43 Scheiner, P. (recorded by Huygens Oeuvres Com. de Christian Huygens, Tome 17me 351–516, La Haye 1932. Also see Bravais, note 38, above).

44 Visser, S.W., 1961, 'Die Halo-Erscheinung', *Handbuck der Geophysik*, **8**, Borntraeger, Berlin, Chapter 15.

45 Whalley, E., 1981, 'Scheiner's halo: Evidence for ice Ic in the atmosphere', *Science*, **211**, 389–390.

46 Whalley, E. and McLaurin, G.E., 1984, 'Refraction halos in the solar system I: Halos from cubic crystals that may occur in atmospheres in the solar system', *Journal of the Optical Society of America* A, **1**, 1166–1170.

47 Weinheimer, A.J. and Knight, C.A., 1987, 'Scheiner's halo: cubic ice or polycrystalline hexagonal ice?', *Journal of Atmospheric Science*, **44**, 3304–3308.

48 Hevelius, J., 1662, '*Mercurius in Sole Visus...*', S. Reiniger.

49 Whipple, F.J.F., 1940, 'How are mock suns produced?', *Quarterly Journal of the Meterological Society*, **66**, 275–279.

50 McDowell, R.S., 1974, 'The formation of parhelia at higher solar elevations', *Journal of Atmospheric Science*, **31**, 1876–1884.

51 Hall, F.F., 1970, 'Observation and explanation of the parhelia 90° from the sun', *Journal of the Optical Society of America*, **60**, 1544.

52 McDowell, R.S., 1979, 'Frequency analysis of the circumzenithal arc: Evidence for the oscillation of ice-crystal plates in the upper atmosphere', *Journal of the Optical Society of America*, **69**, 1119–1122.

53 Tricker, R.A.R., 1971, 'The Kern arc', *Weather*, **26**, 315.

54 Greenler, R.G., Mallmann, A.J., Drinkwine, M., and Blumenthal, G., 1972, 'The origin of sun pillars', *American Science*, **60**, 292–302.

55 Sassen, K., 1980, 'Remote sensing of planar ice crystal fall attitiudes', *Journal of the Meteorological Society.* Japan, **58**, 422–429.

56 Sassen, K., 1980, 'Light pillar climatology', *Weatherwise*, **33**, 259–262.

57 Sassen, K., 1987, 'Polarization and Brewster angle properties of light pillars', *Journal of the Optical Society of America* A, **4**, 570–580.

58 Maunsell, F.G., 1951, 'Parhelic circle', *Weather*, **6**, 245.

59 Lynch, D.K., 1979, 'Polarization models of halo phenomena: The parhelic circle', *Journal of the Optical Society of America*, **69**, 1110–1103.

60 Greenler, R.G. and Mallman, A.J., 1972, 'Circumscribed halos', *Sciences*, **176**, 128–131.

61 Lynch, D.K., Gedzelman, S.D., and Fraser, A.B., 1994, 'Subsuns, Bottlinger's rings and elliptical halos', *Applied Optics*, **33**, 4580–4589.

62 Scorer, R. and Verkaik, A., 1989, p. 158 in *Spacious Skys*, David & Charles, London.

63 Parry, W.E., 1821, *Journal of the Voyages for the Discovery of a Northwest Passage*, Murray, London (reprinted 1968, Greenwood Press, New York).

64 Greenler, R.G., Mallmann, A.J., Mueller, J.R., and Romito, R., 1977, 'Form and origin of the Parry arcs', *Science*, **195**, 360–367.

65 Lynch, D.K. and Schwartz, P., 1979, 'Formation of the anthelion', *Journal of the Optical Society of America*, **69**, 383–386.

66 Mallmann, A.J. and Greenler, R.G., 1979, 'Origins of anthelic arcs, the anthelic pillar, and the anthelion', *Journal of the Optical Society of America*, **69**, 1107–1112.

67 Takano, Y. and Liou, K., 1990, 'Halo phenomena modified by multiple scattering', *Journal of the Optical Society of America*, **7**, 885–889.

68 Greenler, R.G., Mueller, J.R., Hahn, W. and Mallamnn, A.J., 1980, 'The 46° halo and its arcs', *Science*, **206**, 643–649.

69 Lowitz, T., 1794, 'Description d'un meteore remarquable observé a St. Petersbourge le 18 Juin 1790', *Nova Acta Acad. Imp. Sc. Petropol*, **8**, 384–389.

70 Mueller, J.R., Greenler, R.G. and Mallmann, A.J., 1979, 'Arcs of Lowitz', *Journal of the Optical Society of America*, **69**, 1103–1106.

71 Liljequist, G.H., 1956, Halo phenomena and ice crystals, in *Norwegian-British-Swedish Antarctic Expedition 1949–1954, Scientific Results II*, Part 2A, Norsk Polarintitutt, Oslo.

72 Hastings, C.S., 1920, 'A general theory of halos', *Mon. Weather Reviews*, **48**, 322–330.

73 Tricker, R.A.R., 1973, 'A simple theory of certain helical and anthelic halo arcs', *Quarterly Journal of the Royal Meteorological Society*, **99**, 649–656.

74 Wegener, E., 1926, 'Theorie der Haupthalos', *Archive der Deutschen Seewarte*, **43**, no. 2 (Hamburg).

75 Mallmann, A.J. and Greenler, R.G., 1979, 'Origins of anthelic arcs, the anthelic pillar, and the anthelion', *Journal of the Optical Society of America*, **69**, 1107–1112.

76 Greenler, R.G. and Trankle, E., 1984, 'Anthelic arcs from airborne ice crystals', *Nature*, **311**, 339–343.

77 Whymper, E., 1871, *Scrambles Amongst the Alps*. John Murray, London.

78 Trankle, E. and Greenler, R.G., 1987, 'Multiple scattering effects in halo phenomena', *Journal of the Optical Society of America* A, **4**, 591–599.

79 Takano, Y. and Liou, K., 1990, 'Halo phenomena modified by multiple scattering', *Journal of the Optical Society of America* A, **7**, 885–889.

80 Ripley, E.A. and Saugier, B., 1971, 'Photometeors at Saskatoon on 3 December 1970', *Weather*, **26**, 150–157.

81 Evans, W.F.J. and Tricker, R.A.R., 1972, 'Unusual arcs in the Saskatoon display', *Weather*, **27**, 234–236.

82 Takano, Y. and Asano, S., 1983, 'Franhofer diffraction by ice crystals suspended in the atmosphere', *Journal of the Meteorological Society of Japan*, **61**, 289–300.

83 Broomall, C.M., 1901, 'Lunar halo', *Sciences*, **13**, 549.

84 Hakumaki, J. and Pekkola, M., 1989, 'Rare vertically elliptical halos', *Weather*, **44**, 466–473.

85 Pekkola, M., 1988, 'Arvoituksellinen Hissikin halo', *Tahdet ja Avaruus*, **18**, 60–63.

86 Schlesinger, F., 1913, 'Elliptical lunar halos', *Nature*, **91**, 110–111.

87 Pekkola, M., 1991, 'Finnish Halo observing network: Search for rare halo phenomena', *Applied Optics*, **30**, 3542–3544.

88 Riikonen, M. and Ruoskanen, J., 1994, 'Observations of vertically elliptical halos,' *Applied Optics*, **33**, 4537–4538.

89 Parviainen, P., Bohren, C.F. and Mäkelä, V., 1994, 'Vertical elliptical coronas caused by pollen', *Applied Optics*, **33**, 4548–4551.

90 Gadsden, M. and Schröder, W., 1989, *Noctilucent Clouds*, Springer-Verlag, Berlin.

6 Naked-eye astronomy

Imagine a comet ten times brighter than the full moon and plainly visible in daytime. That was the case of the sun-grazer Ikeya–Seki as it approached the solar disk mid-day on October 21, 1965. Torn by powerful tidal forces in the proximity of the sun, its tail began to lag far behind the sun–comet line and assumed the shape of a fish hook. Shortly afterward the head was observed to disintegrate into three parts.

DAYTIME

6.1 The sun

The sun is our star. It gives light and heat to the solar system. All the planets, comets, meteoroids and interplanetary dust orbit the sun[1]. Without it to light our world, few of the phenomena in this book would exist.

The sun is an incandescent ball of gas with a surface temperature of about 5400 °C. It is 109.3 times the diameter of the earth and 333 400 times as massive. The mean distance to the center of the sun is 149 598 000 kilometers, and this is called the astronomical unit or AU. Since the earth's orbit is slightly elliptical, the distance varies by a few percent over the year. Perihelion, or the closest approach, occurs on January 4 when the earth is 147 000 000 kilometers away. Aphelion, or farthest recession, happens about six months later when the earth is 152 000 000 kilometers away. This eccentricity in the orbit makes the solar disk change size in the sky from 32.5 arc minutes in January to 31.3 arc minutes in July. The mean angular diameter is about 32 arc minutes or just over 0.5°. We become aware of variability of the sun's diameter in connection with total eclipses of the sun where even a 1 arc second change would affect the geographic width of the region in which totality is observed.

The earth's orbital motion makes the sun appear to move eastward in the sky with respect to the stars by about 1° per day, or twice its own diameter. At the equinoxes (March 21 and September 21) the sun is also moving north or south at a rate of

0.4° per day, or slightly less than its diameter. At the summer solstice (June 21) it reaches its furthest point north at a declination +23.5° after which it begins to move south. On the autumnal equinox it crosses the celestial equator traveling south and attains its most southern point at the winter solstice (December 21).

Composed entirely of gas the sun has no solid surface. But just as a fog bank can appear sharply bounded from a distance, the sun has an apparent surface called the photosphere. By analogy with fog, if we could approach closely enough we would begin to discern that its edge is diffuse. This transition spans a distance so small that from the earth the photosphere appears sharp even with a large telescope.

If the sun is viewed through low-lying clouds, thin fog, or smoke, we can readily discern this photospheric disk and we can see that it is not uniformly bright but displays 'limb darkening'. In other words there is a noticeable decrease of brightness toward the edge of the disk. Limb darkening is a consequence of the solar atmosphere being cooler with height.

6.2 Sunspots

Small dark areas on the solar disk are called sunspots (Figure 6.2). Galileo officially discovered sunspots with his telescope in 1611, but since they can easily be seen with the naked eye under the right conditions many people must have noticed them before. Spots as small as 1 arc minute across can be made out and if we are persistent, they can be seen far more frequently than is commonly supposed, especially during those years when their numbers are greatest[2-5]. They are best viewed through thin fog or with the aid of a dark filter or overexposed black and white film. In humid regions of the world, prevalent low-lying haze causes the rising solar disk to be extremely reddened and dim, rendering large sunspots visible almost daily. With a long 'focal length' pinhole camera we can detect smaller spots than otherwise (see page 18 in Giovanelli[6]).

Sunspots are transient events lasting a few weeks or months. A large sunspot observed on several days will move across the disk from east to west. This is a consequence of solar rotation which has a period of about 27 days. Sunspots are dark because they are cooler than the surrounding photosphere and emit less light.

The number of sunspots on the disk varies with the sunspot cycle[7]. On the average the interval between the successive maxima in numbers is 11.1 years, although each cycle is somewhat different in length. Practically no sunspots are present at solar minimum, while at maximum there are usually many. Table 6.2 lists the times of minimum and maximum for the past 150 years and has been extrapolated, with some uncertainty, into the future.

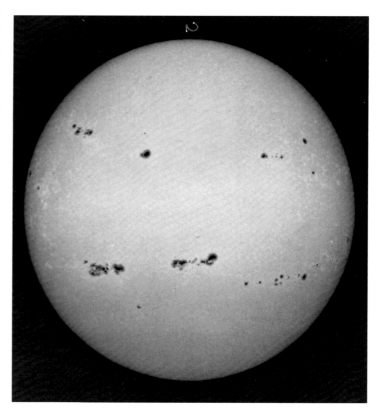

Fig. 6.2 The solar disk near sunspot maximum on November 9, 1979. The larger of these sunspots would be visible to the unaided eye. Notice that the spots concentrate in two parallel bands, north and south of the sun's equator. The shading toward the edge of the disk is called limb darkening.

Table 6.2. *Dates of sunspot minimum and maximum.*

Cycle	Minimum	Maximum
8	1833	1837
9	1843	1848
10	1855	1860
11	1867	1870
12	1878	1883
13	1890	1894
14	1902	1906
15	1913	1917
16	1923	1928
17	1933	1937
18	1944	1947
19	1954	1958
20	1964	1968
21	1976	1979
22	1986	1989
23	1996	2001
24	(2006)	

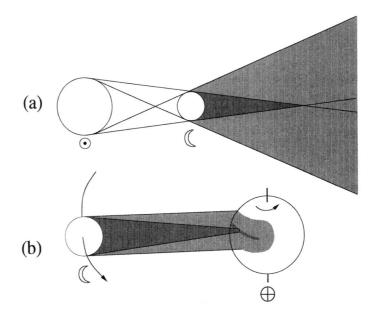

Fig. 6.3A Circumstances of a solar eclipse. Not to scale (a) the usual umbral–penumbral shadow of the moon. Where the umbra intersects the earth we have a total eclipse; the penumbra produces a partial eclipse (b).

6.3 Solar eclipses

By remarkable coincidence the apparent diameters of the sun and moon are nearly the same. Depending on position in their respective elliptical orbits, the size of the moon varies between 90.2% and 106% of the sun's angular diameter.

When the moon passes in front of the sun's disk as viewed from the earth we have a solar eclipse. If the moon's umbral shadow intercepts the earth's surface at a time when the moon appears larger than the sun we have a total eclipse and the faint corona becomes visible (Figure 6.3A). If the moon is smaller we have an annular eclipse. In this case a bright ring of the photosphere remains visible and there is no corona. Within the shadow's penumbra it is a partial eclipse. On average, the width of totality on the ground is only 100 kilometers, while the area where it is seen partial may extend over 6400 kilometers. There can be up to five solar eclipses per year, although two is the likely number. On average, a total eclipse will repeat itself at a given location once in 375 years; an annular one once in 224 years. So the latter event is slightly more common at a given place[8].

The moon's shadow sweeps over the earth at about 3380 kilometers per hour. Because the earth is rotating in the same direction, this helps to slow down the shadow to 1270 kilometers per hour at the equator. On July 11, 1991 the weather satellite GOES 7 photographed the shadow as it approached Baja California, thus recording what we could only imagine before (Figure 6.3B).

So exactly are the orbits of the sun and moon known, and consequently the areas on the earth's surface where totality takes place are so well determined geographically, that historic references to eclipses can provide clues as to whether the sun may have changed in size with time. A difference of 1 arc second in the diameter of the sun can be deduced from the simple historical observation as to whether or not, during the eclipse of 1715, the corona was seen in the English village of Cranbrook in the south and also at Darrington in the north along the path of totality[9]. Scientists are now delving into such past references to eclipse sightings in order to learn if the sun has varied in size (a factor related to understanding climate change).

Fig. 6.3B Mosaic of photographs taken from a weather satellite of the moon's shadow projected onto the earth's surface on July 11, 1991. The actual umbra area lies within this black blotch and cannot be made out. The penumbra boundary extends over much of this print but is too faint to see. (NASA/NOAA)

6.4 Total solar eclipse phenomena

A total eclipse of the sun rates along with volcanic explosions and tornados as the most spectacular sights in nature. No effort should be spared to witness at least one total eclipse in your lifetime[10-11].

An eclipse begins with little fanfare. At 'first contact' the invisible new moon takes its first bite out of the sun. Except for the disk becoming occulted, nothing very striking happens until about 10 minutes before totality when we become aware that the sunlight is failing. Several unfamiliar conditions then ensue. For one, the sky begins to darken without the accustomed reddening that precedes sunset. Hues that are normal for mid-day continue to prevail even though it is getting dark. Shadows become strange. Openings in tree foliage overhead, which normally cast circular patches of light on the ground, now project crescent shapes (Figure 6.4A). If we examine the shadows of our own hands, finger sharpness depends on the angle we hold them (Figure 6.4B). This is a consequence of the sun becoming a slit-like source of light, broad in one direction, narrow in the perpendicular. Within 5 minutes of totality shadow bands may begin running across the ground. These shadow bands are low contrast, roughly parallel, ridges of light and dark that would escape our attention except for their movement. The distance between the bands varies from 5 centimeters to perhaps 2 meters; their velocity is 2–20 meters per second. Attempts to photograph them have been largely unsuccessful[12] but the eye is especially good at seeing patterns of this kind. These bands arise from air density fluctuations high in our atmosphere; the same reason that stars twinkle.

About 30 seconds before totality one has the the growing sense that something is desperately 'wrong'. The sky itself now darkens rapidly as the shadow of the moon sweeps toward us (Figure 6.4C).

Heralding the onset of totality is the Diamond Ring (Figure 6.4D). The last sliver of photospheric light combines with the emerging ring of the inner corona to mimic a diamond ring in the sky. The ring part is sometimes peppered by Baily's Beads, sparkling points of photospheric light shining through lunar valleys.

Suddenly totality is upon us. This is the moment of second contact when the moon first completely covers the sun and the solar corona is now revealed in all its splendor! What you are witnessing can be seen in no other way and is unique to this moment. Neither spacecraft nor the most powerful ground-based telescopes can record what your own eyes now behold. Even photographs taken

Fig. 6.4A Dappled sunlight cast on the ground by gaps in the foliage are normally circular, being in fact 'pinhole' images of the sun. During the partial phase of an eclipse these patterns become crescent shaped; in the case of an annular eclipse they would be ring shaped.

Fig. 6.4B Near second and third contact, shadows of our hands sharpen in one direction as here demonstrated at the Hyderabad eclipse camp, February 16, 1980.

by astronomers at this same eclipse will not do justice to your personal perception of this visual scene. For once the unaided eye reigns supreme.

The corona is a soft, pearly white light encircling the eclipsed sun. Structures in the form of helmets, fans, and rays can be traced out to one or more solar diameters. The corona is the tenuous outer atmosphere of the sun at a temperature of 1–2 million degrees. This makes it the hottest object in the universe visible to the naked eye. During sunspot maximum, the inner parts of the corona are fairly symmetrical and as bright as the full moon, or one millionth the intensity of the photosphere (Figure 6.4E). Around sunspot minimum, the coronal equatorial rays are unsymmetrical and develop to their greatest elongation[13].

The corona is polarized. This is because sunlight is scattered off free electrons. It is tangentially polarized up to 40% at a distance outward of one quarter the radius of the sun from its edge. Beyond about one solar radius, interplanetary dust particles begin to

Fig. 6.4C Shadow of the moon darkens the horizon as seen from the summit of Mauna Kea, Hawaii, on the eclipse morning of July 11, 1991.

Fig. 6.4D Diamond Ring as seen at second contact from Hyderabad, India, February 16, 1980.

dominate over electrons as the scattering source and polarization diminishes. These are the same particles responsible for the zodiacal light (Section 6.18).

Close to the occulting moon we may see tiny, red features called prominences (Figure 6.4F). Their color originates primarily from a spectral line of hydrogen in emission at 656 nanometers. At both the second and third contacts the chromosphere flashes out briefly as a narrow reddish arc. This is the hot, normally invisible layer overlying the photosphere. Again the red light of hydrogen dominates. Binoculars greatly assist in seeing the chromosphere and prominences.

A few bright stars and planets can be seen during totality. If one is located on a hill, a view towards the horizon may show the edge of the moon's umbra against the still illuminated distant atmosphere. One could watch the progression of this shadow across the landscape, but this observation would cost precious time that is best devoted to the corona.

At third contact the events preceding second contact take place in reverse order: there is the chromospheric flash, Baily's beads, and then the blossoming diamond ring drives the corona to invisibility. Before you know it, the spectacle is over. By fourth contact, when the moon departs the disk, most expeditions are packing and making ready to return home.

Fig. 6.4F Prominences show up as distinctive red features around the moon's disk.

Fig. 6.4E Solar corona typical of activity conditions mid-way between sunspot number minimum and maximum. Observed with a graded filter on June 30, 1973, from Lake Rudolph in Kenya (High Altitude Observatory photo). The graded filter positioned in front of the film helps to equalize the intensity of the inner and outer corona, a difference the unaided eye easily accommodates.

6.5 Eclipse danger to the eye

The eclipse observer should be aware that the retina of the eye can be permanently damaged during the partial phase near second and third contact unless care is exercised. Ordinarily if we look at the sun directly its extreme brightness causes us to turn away reflexively. This is a natural form of protection because the solar image, if held fixed on the retina for several seconds, would destroy the local tissues by heating. During the partial phases, when the moon has covered most of the solar disk, the total brightness of the sun is much reduced and this inherited reflexive protection fails; the iris dilates. Even so, the energy per square centimeter in the crescent as projected onto the retina remains the same as for outside eclipse. Unaccompanied by any sensation of discomfort, a prolonged stare at this nearly eclipsed sun will scar your retina.

With the above in mind, the rule is never to more than glance at the crescent phase. If you wish to study the partially eclipsed sun either examine a projected image on a card, as one does outside eclipse, or look at the sun through overexposed black and white photographic film. During totality it is completely safe to look at the corona without protection.

6.6 Unusual eclipse phenomena

The following is a list of possible eclipse-related phenomena to watch for near totality. You only have a few minutes and a prepared mind will help one to get the most out of the experience. Some are subtle[14]; all have been reported but few photographed.

- Coronal transients (short-lived, bubble-like enhancements in coronal structure).
- Comets near the sun (which are invisible outside eclipse).
- Edge of the moon opposite sun is visible in partial phase (indicating the existence of coronal light which would otherwise go unnoticed).
- Inner parts of the corona seen before totality.
- Brighter stars and planets seen outside totality.
- Prominences seen several minutes before or after totality.
- Moon darker than surrounding sky during totality (Mach effect? Section 7.7).
- Earthshine visible on the eclipsed moon[15]; look for lunar features.

ECLIPSE PHOTOGRAPHY

Our advice is: don't bother! You have traveled halfway around the world and these few moments of totality are precious. Why screw around with cameras, lenses, lens caps, cable releases, and a dozen other distractions? Let the pros do the job; they have better equipment and will be happy to share their results. Instead sit back in a comfortable lawn chair with your binoculars and just enjoy it.

If you cannot resist taking pictures (and we do understand), then here are a few tips. The brightness of the corona is about the same as a full moon which your normal landscape exposure would capture. Ignore your exposure meter and set for this condition and bracket that setting. Under expose and you will record prominences and the bright inner corona. Over expose, but not too much, and you will see the corona further out. Extensive instructions on all aspects of eclipse photography can be found in the relevant NASA publications and in various popular books on eclipses.

6.7 Annular and partial eclipses

When the lunar disk is smaller than the sun's, we can have an annular eclipse at true opposition. Sometimes called a 'ring' eclipse, part of the photosphere remains visible throughout and its glare prevents our seeing the corona, chromosphere, and prominences (Figure 6.7). Many of the shadow phenomena and lighting effects described above will still transpire and the event has its own attractions. Baily's beads are sometimes reported. If the eclipse is central, sunlight shining through tree leaves casts doughnut-shaped images of the sun on the ground.

Partial eclipses are widely observed and relatively common. It is perfectly possible for a partial eclipse to cover much of Africa and South America in a single pass. The fraction of the sun eclipsed depends on your distance from the mid-eclipse line. As the moon appears to move eastwards in the sky about 13 times as fast as the sun, it will cross the disk from west to east. The time span is again determined by your distance from central eclipse.

Fig. 6.7 Annular or 'ring' eclipse observed at San Diego, California, on January 4, 1992[16].

6.8 Future solar eclipses

The Astronomical Almanac[17], issued annually by the US Naval and Royal Greenwich Observatories, contains fairly complete information on both solar and lunar eclipses each year (Figure 6.8). Table 6.8 lists the total eclipses from 1990 to 2015.

Fig. 6.8 Circumstances for the eclipse of July 11, 1991. The path of totality, the two bold parallel lines, passed directly over the world's largest astronomical observatory on Mauna Kea, Hawaii, the tip of Baja California (a dry desert), and into Mexico. The boundaries of the penumbral shadow (partial phase) extended over much of the Western Hemisphere. Times are in UT.

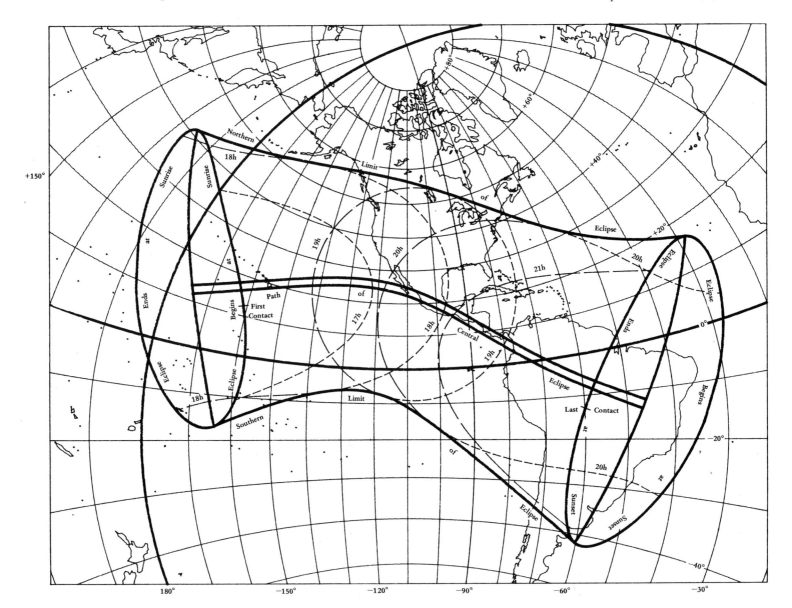

Table 6.8. *Total and annular solar eclipses*[7].

Date	Duration of totality (m:s)	Type	Path
1990 Jan 26	2:06	annular	S Atlantic, Antarctica
1990 Jul 22	2:33	total	Finland, Novaya Zemlya, N Siberia, Pacific
1991 Jan 15	7:55	annular	SW Australia, Tasmania, New Zealand, Pacific
1991 Jul 11	6:54	total	Pacific, Hawaii, Mexico, Central & South America
1992 Jan 4	11:42	annular	Pacific
1992 Jun 30	5:20	total	Atlantic
1994 May 10	6:14	annular	Pacific, USA, Atlantic
1994 Nov 4	4:23	total	South America, S Atlantic
1995 Apr 29	6:30	annular	Pacific, South America
1995 Oct 29	2:10	total	Asia, Borneo, Pacific
1997 Mar 9	2:50	total	Siberia
1998 Feb 26	4:08	total	Pacific, Central America, Atlantic
1998 Aug 22	3:14	annular	Sumatra, Malaysia, Borneo, New Guinea, Pacific
1999 Feb 16	1:19	annular	Australia, Indian Ocean
1999 Aug 11	2:23	total	Atlantic, France, Middle East, India
2001 Jun 21	4:56	total	Atlantic, S Africa, Madagascar
2001 Dec 14	3:54	annular	Pacific, Central America
2002 Jun 10	1:13	annular	Pacific
2002 Dec 4	2:04	total	S Africa, Indian Ocean, Australia
2003 May 31	3:37	annular	Iceland
2003 Nov 23	1:57	annular	Antarctica
2005 Apr 8	0:42	annular	Pacific, Central America
2005 Oct 5	4:32	annular	Atlantic, Spain, Africa, Indian Ocean
2006 Mar 29	4:07	total	Atlantic, Africa, Turkey, Russia
2006 Sep 22	7:09	annular	NE South America, Atlantic, Indian Ocean
2008 Feb 6	2:14	annular	S Pacific Ocean, Antarctica
2008 Aug 1	2:28	total	N Canada, Arctic Ocean, Siberia, China
2009 Jan 26	7:56	annular	S Atlantic, Indian Ocean, Sumatra, Borneo
2009 Jul 22	6:40	total	Asia, Pacific
2010 Jan 15	11:11	annular	Africa, Indian Ocean, SE Asia
2010 Jul 11	5:20	total	Pacific, S South America
2012 May 20	5:47	annular	China, Japan, N Pacific, USA
2012 Nov 13	4:02	total	N Australia, Pacific
2013 May 10	6:04	annular	Australia, Pacific
2015 Mar 20	2:47	total	N Atlantic, Svalbard, Arctic Ocean

NIGHTTIME

6.9 Artificial satellites

On any clear night man-made satellites, spent boosters and other space debris can be seen drifting among the stars (Figure 6.9A). They are visible by reflected sunlight and look like stars except that they move. Most artificial satellites travel across the sky in a west to east direction, although they also have a north or south component to their motion. They are normally launched in the west–east direction to take advantage of the earth's rotation which gives them a running start into orbit. Polar orbiters travel primarily north and south.

The majority of satellites are in a low earth orbit between 185 and 555 kilometers (Figure 6.9B). Their orbital period is about 90 minutes, and they appear to move in the sky at several degrees per second; their linear velocity is 8.5 kilometers per second. In the early evening, a satellite will slowly fade as it enters the earth's shadow. When viewed with binoculars the brighter ones turn red as they catch the last rays of the setting sun[19]. Could we observe a satellite crossing the solar disk? A calculation shows its transit time would be about 1.3 seconds and its diameter approximately 10 arc seconds. Not much chance of seeing that!

Because their orbits are close to the earth, satellites are only visible for a few minutes before going over the observer's horizon. People resident at higher latitudes seldom see the large payloads which are boosted into orbit from Cape Canaveral because these are usually below the horizon. Indeed, since the orbital inclination of large US satellites is precisely equal to the latitude of the Kennedy Space Center, people more than 15° beyond latitude 28.5° can never see them. The satellites they observe are generally Soviet ones, launched from near 65°, or are polar orbiters[20].

Some satellites, especially communication satellites, are in geosynchronous orbits at 35800 km. Their orbital periods are precisely 24 hours so that they hover over the same spot on earth (Figure 6.9C). They spend most of their time in direct sunlight but, with astronomical magnitudes between 11 and 14, these 'geostats' are normally too

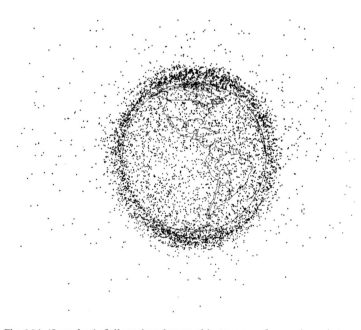

Fig. 6.9A 'Snapshot' of all cataloged space objects as seen from a viewpoint over the western hemisphere at a particular instant on January 1, 1986. (After Maley[22])

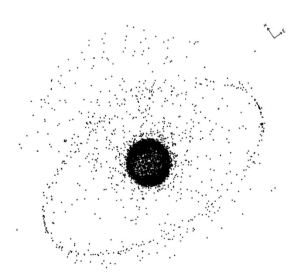

Fig. 6.9B 'Snapshot' of all space objects on the same occasion but from a more remote viewpoint than in Figure 6.9A so as to show the outermost geosynchronous satellites and other space debris. The concentration in the north represents in large part Soviet Molniya orbit satellites.

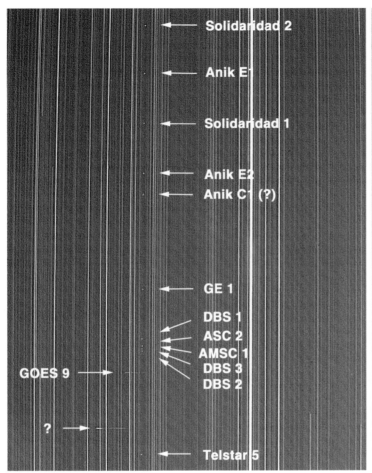

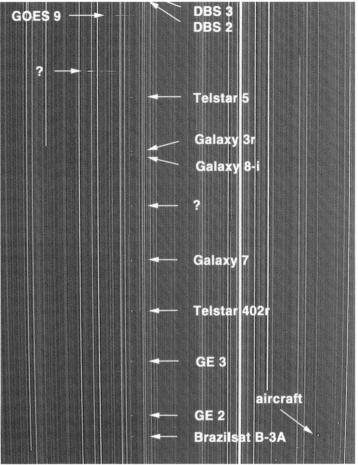

Fig. 6.9C Geostationary satellites from Kitt Peak (longitude: 111.5° west, latitude: +31.95° north) on May 16, 1999. This is an 8-hour exposure at f/6.3.

faint to be seen with the naked eye[21]. Even so, an occasional specular reflection from their solar panels causes a momentary glint[22]. This is the probable cause of the 'Aries Flasher', reports of which provoked wide speculation as to its origin beginning in the summer of 1984[23].

Because geostats do not move in the sky they can be photographed by a fixed camera. All that is needed is a 35 millimeter camera and a tripod. Satellites as faint as 12th magnitude can be recorded[24,25].

Those few US satellites in polar orbits (north–south motion) and many Soviet satellites have 12 hour highly eccentric *Molniya orbits*, causing them first to hang nearly motionless over high latitudes (see Figure 6.9B) for several hours near apogee, or most distant from the earth position, then to swoop rapidly across the sky near perigee, or closest.

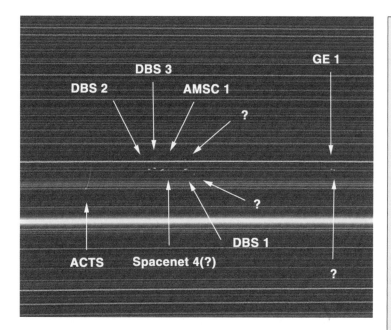

Fig. 6.9D Detail from a 10 hour exposure on March 19, 1999. Imperfections of satellite orbital motions show as slight elongations. DBS signifies Direct Broadcast Service and may be the origin of your home TV signals. ACTS is now an abandoned Israeli satellite. AMSC represents American Mobile Sat. Corp.; GE is General Electric, etc. Unidentified objects, which may be secret, are labeled '?'.

HOW TO PHOTOGRAPH GEOSTATIONARY SATELLITES

The following is a proven recipe for the photography of geostationary objects:

(1) Conditions: Clear sky, no moon, no clouds, no city lights.

(2) Use a sturdy tripod or, even better, place your camera directly on rocks on the ground. Any camera movement from wind, settling, or thermal expansion/contraction is fatal. Propping the camera with rocks is best in our experience.

(3) Point your camera due south (northern hemisphere) on the meridian and at an altitude of $(90° - \phi)$, where ϕ is latitude. Except at the equator one must compensate for a latitude-dependent parallax as follows[26]:

$\phi°$	corr. (°)	$\phi°$	corr. (°)
0	0	50	7.3
10	1.8	60	8.1
20	3.5	70	8.5
30	5.0	80	8.7
40	6.3		

We use a machinist's protractor with a bubble level to set the elevation. The south direction can be found with the help of a magnetic compass, taking account of the local magnetic deviation which is given on most topographic maps.

(4) Film: Ektachrome 100S, an emulsion with low reciprocity failure.

(5) Exposure: 6 to 8 hours at f/8 with Ektachrome 100S film. (We like the long exposure because satellite motions are revealed and one can get uninterrupted sleep. Shorter exposures such as 4 hours at f/5.6, 2 hours at f/4, etc., may be prefered if clouds or dew are likely).

Proceed based on your results. If you don't reach sky background you will not record satellites. Over exposure is bad too. If Ektachrome 100S is unavailable one can compensate somewhat by making a slight twilight pre-exposure and/or extra development (request 'push two stops' when you take the film in for processing, for example). We have used an 80 mm f/2.8 Hasselblad, but other cameras work as well. Remember, these are faint 12th magnitude objects.

6.10 The moon

The earth's natural satellite is the moon. Besides being a handsome object, it helps us to see at night. A newspaper can be read by the full moon. At this phase it has an intensity about one millionth of the sun.

Although it looks bright and colorless, in fact it is about the brightness and color of brown soil (Figure 6.10). During the day, when the moon is high in the sky (to avoid bluish airlight), hold a white piece of paper up and compare it to the moon's disk. The moon will be fainter than the paper. This relatively dark object has an albedo, or reflectivity, of about 7%, which is among the lowest in the solar system (Section 6.14).

The dark areas on the moon's disk are called maria, or 'seas', because early observers assumed the moon was much like the earth. The contrast between the maria and the brighter parts, the mountains, is about 20%. Maria are dark because they are volcanic basalt flows, mostly of iron silicate. The mountainous 'highlands' consist of igneous crustal deposits of aluminum and calcium silicates.

Fig. 6.10 Color photograph, by Charles Hunter, of the moon displaying its natural brownish hue. Airlight in the daytime and dazzle at night prevent this perception.

Passed from generation to generation by word of mouth, every society has a favorite imagined figure that is seen in the surface markings of the full moon[27]. In the Orient and in Europe it is commonly a hare. To North Americans it is the 'Lady in the moon'. At moonrise she is seen in profile, her hair in the style of the 1890s Gibson girl, looking downward[28]. At moonset, her gaze is upwards. Familiarity with this lady allows one to decide on inspecting a photograph containing a full moon whether it was taken looking east or west.

The moon keeps the same face turned towards the earth at all times. Tidal distortion by the earth over past eons has slowed its rotation period to match its orbital period. This does not mean, however, that we only see half the moon's surface from the earth. The eccentricity of the lunar orbit and our parallactic displacement as the earth rotates allows one to see additional lunar surface. These oscillation-like movements, called librations, expose an extra 18% of the moon's surface at one time or another. Although librations went unnoticed until Galileo's telescopic discoveries, systematic naked-eye mapping of the moon can reveal the effect[29]. It seems that the capability of the eye was not used to full advantage for study of the moon before the invention of the telescope. Why this was so is not understood[30].

Orbital motion eastwards causes the moon to rise an average of about 50 minutes later each 24 hours. In one day the moon moves about 12° along its orbit (with respect to the stars) and the earth must rotate further to bring it back to its original sky elevation. After one lunar month (about 29 days relative to the sun) the moon has completed one revolution of the earth (29.5306 days × 12.191° per day = 360°).

A harvest moon rising in the northeast is the nearly full moon near the time of the autumnal equinox in September. Owing to the inclination of the lunar orbit to the horizon, the moon rises 22 minutes later (plus 24 hours) for several nights in succession, instead of the usual 50 minutes, thereby providing extra light coincidentally with the autumn harvest. Six months later, near the vernal equinox, the debt is repaid and the time between successive moon rises may be almost one and a half hours.

Many stars and planets are regularly occulted by the moon. Such occultations provide valuable scientific information about lunar, stellar, and planetary positions, as well as data on planetary atmospheres, the existence of double stars, and stellar diameter measurements.

6.11 Phases of the moon

The moon goes through phases because it orbits the earth in nearly the same plane as the earth orbits the sun. By holding a tennis ball at arm's length when the sun is low and turning yourself around, you will see the tennis ball go through similar phases.

An exactly new moon is invisible because it is in conjunction with the sun and only the dark side faces us. A few nights later a thin crescent hangs in the evening twilight. At this time the dark side of the moon can be faintly seen illuminated by earthshine (Section 6.12).

Earliest detection of the new crescent moon is a matter of commercial importance in the Islamic world and is the object of a longstanding competition among amateur astronomers. For the lunar month to begin a Muslim cleric must sight the crescent and declare its existence. This has led to jet aircraft being sent aloft for this purpose.

In the western world it is simply the honor of taking the first possible photograph[31]. Any sighting less than 24 hours after a new moon is noteworthy, with the record being 14.5 hours[32]. The French astronomer André Danjon investigated the limit in some detail. He found that were the moon smooth like a billiard ball it would be visible practically up to the moment of new moon, but the roughness of the lunar mountains cast shadows, and these impose a definite cutoff at 7° from the sun. This 7° is more than the maximum distance (5.5°) of the moon from the sun at new phase, and so the lunar crescent must always vanish for some period around new moon. To improve your chance of seeing a young crescent, observe at the vernal equinox when the motion of the moon is greatest so that it travels further in a given time. Also viewing the moon at perigee helps because, for a given interval past new moon, it is further from the sun.

When the moon forms a right angle with the earth–sun line (called quadrature), it is at first quarter and the 'half full' moon is on the meridian at sunset. Between first quarter and full the phase is the waxing gibbous. When the sun, earth and moon are roughly aligned (in opposition) we have a full moon and the setting sun is then balanced by its rising disk. A lunar twilight, or glow in the

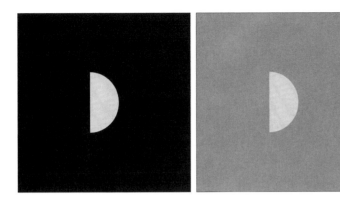

Fig. 6.11 The two semicircles representing the moon are the same brightness. Yet the one on the left appears brighter than the one on the right because of different background brightness.

Table 6.11 *Relative illuminance on a horizontal surface from the sun, twilight sky, and full/quarter moon.*

Altitude (deg)	Horizontal illumination (lumen/m^2)		
	Sun	Full moon	Quarter moon
90	1 000 000	2.00	(0.180)
55	643 700	1.80	0.158
40	420 700	1.17	0.104
30	309 600	0.815	0.073
20	185 000	0.500	0.040
10	77 000	0.160	0.0020
5	37 000		
1	7 000		
0	4 000		
−1	2 000		
−4	110		
−6	12	civil twilight ends	
−7	4		
−8	1.5		
−10	0.23		
−12		nautical twilight ends	
−16	0.013		
−18		astronomical twilight ends	

eastern sky, will precede moonrise when the phase is just past full. After full moon the phases run through waning gibbous, third quarter, and finally the sliver of a crescent slides into morning twilight, again accompanied by earthshine.

The full moon is roughly ten times brighter than the quarter phase even though there is only a factor of two difference in illuminated area. As discussed in Section 1.5, this brightness enhancement comes from the opposition effect. Table 6.11 shows how lunar and solar landscape illuminance compare.

6.12 Earthshine

The faint light seen on the dark side of the moon when it is a thin crescent is earthshine (Figure 6.12). It represents sunlight reflected from the earth to the moon and is variable because the earth's reflectivity changes as large cloud masses come and go. In a sense the moon with its earthshine acts like a crude weather satellite by reporting on the general state of terrestrial cloudiness[31].

As the phase of the moon progresses beyond new, earthshine fades in a day or two. There are three reasons for this rapid fall off of earthshine. First, the amount of light from the earth decreases. If the earth could be seen from the new moon, it would appear in full phase. As the moon moves around the earth, the earth enters a waning gibbous phase and therefore the amount of sunlit earth available to make earthshine diminishes. Second, the opposition brightening of the earth is lost. Finally, there is the increasing glare of the waxing crescent which causes a loss of visibility by irradiation (Section 7.8).

Fig. 6.12 Earthshine allows us to see the 'dark' side of the moon. It is sunlight reflected first from the earth to the moon, then back again. Because the amount of light reflected from the earth depends on the amount of cloud cover, the brightness of the 'dark' side of the moon varies from month to month. (Photo by D.J. Gutierrez and G. Gutierrez)

6.13 Lunar eclipses

If the moon orbits into the earth's shadow a lunar eclipse takes place. In a partial lunar eclipse the moon passes through the penumbra or only a part of the umbra. In a total lunar eclipse it is completely within the umbra.

Negligible dimming results during the penumbral phase and partial eclipses easily escape our attention. As the moon enters the earth's umbra a darkening encroaches on its eastern limb which steadily marches across its face. Totality marks the moon being completely within the earth's umbra. It can last a maximum of 100 minutes if the moon passes through the center of the umbra.

During some total eclipses, the moon is so dark as to be almost invisible. At other times it is a surprisingly bright, coppery red. The brightness of the moon during totality depends on the amount of light our atmosphere scatters into the umbra. When the earth's terminator is relatively cloud-free, sunlight is refracted through the clear air and onto the moon. This light is red for the same reason that the setting sun is red. And since the light propagates tangentially through the earth's atmosphere, it is reddened twice as much (Figure 6.13). At times the shadow has been reported to be mis-shapened following volcanic eruptions[33]. Presumably the cause is uneven obscuration along the earth's terminator. Imagine how beautiful the event must look from the moon during such an eclipse: the reddish halo of the earth (which subtends 1°.80, or almost four times the diameter of the moon as seen from the earth) surrounded by white coronal rays and its petal-like streamers.

Total lunar eclipses are less common than total solar eclipses because lunar totality implies a complete coverage of the moon (whereas solar totality requires only that the umbra touches down at one point on the earth's surface). There can be as few as zero and as many as three in a given year. Even though lunar eclipses are less frequent, they are visible from half the entire world at one time and so are widely observed.

The magnitude of a lunar eclipse is specified by the ratio of moon's diameter to the width of the earth's umbra cross-section at the moon. A magnitude of 1.0 or greater is total and less than 0.7 proves invisible from a lack of contrast. Details about the current year's eclipses can be found in the *Astronomical Almanac*. Table 6.13 gives the dates and Universal Time (UT) (mid-eclipse) of forthcoming lunar eclipses having perceptible darkening (magnitude greater than 0.7). See Section 7.16 for how to convert UT to your local time. Magnitude is the ratio of umbral shadow to the moon's diameter.

Fig. 6.13 Total eclipse of the moon during the partial phase, March 24, 1975. (Photo by Robert Chambers)

Table 6.13. *Lunar eclipses*[7].

Date	UT	Magn.
Apr 4, 1996	00:11	1.379
Sep 26. 1996	02:55	1.240
Mar 24, 1997	04:41	0.918
Sep 16, 1997	18:47	1.190
Mar 13, 1998	04:22	0.707
Sep 6, 1998	11:11	0.813
Jan 31, 1999	16:20	1.005
Jan 21, 2000	04:45	1.325
Jul 16, 2000	13:57	1.769
Jan 9, 2001	20:22	1.188
Dec 30, 2001	10:30	0.893
Nov 20, 2002	01:47	0.860
May 16, 2003	03:41	1.128
Nov 9, 2003	01:20	1.017
May 4, 2004	20:32	1.303
Oct 28, 2004	03:05	1.307
Apr 24, 2005	09:57	0.865
Mar 3, 2007	23:22	1.233
Aug 28, 2007	10:38	1.477
Feb 21, 2008	03:27	1.106
Aug 16, 2008	21:11	0.806
Feb 9, 2009	14:39	0.899
Dec 21, 2010	08:18	1.257

6.14 The planets

There are five planets bright enough to be visible to the naked eye: Mercury, Venus, Mars, Jupiter, and Saturn (Table 6.14). Uranus at magnitude 5.5 is too near the limit of vision to be reliably found without a telescope. Exceptionally bright 'stars' at dawn or dusk are likely to be planets. Chance conjunctions of the planets create striking displays in the twilight skies[7].

Unlike stars, which are point sources, planets subtend a measurable angle in the sky. Venus can be as large as 64 arc seconds at its closest approach to the earth. As a result of their finite size, planets 'twinkle' less than stars. This can provide a useful test when trying to determine if that bright star in the sky is a planet. Another way is to stand in the shadow of a tall building or distant mountain and watch to see how long it takes for the object to eclipse. If it goes out instantly it is a star. Jupiter takes about three seconds to be extinguished[34].

At its greatest elongation, i.e. its greatest angular distance from the sun, Mercury is −1.8 magnitudes. It is always within 28° of the sun and, being in the twilight sky, is easily missed. Copernicus, the father of our modern view of the solar system, is said to have never seen Mercury. Having the shortest synodic period of any planet (115.9 days), it swings from evening to morning twilight every couple of months.

Venus is the next planet outwards from the sun. Like Mercury, its orbit lies inside the earth's and therefore it also is confined near the sun with never more than a 47° elongation. It is bright enough (up to −4.3 magnitude at an elongation of 40°) to be seen in the daytime if one knows where to look. Often called the Evening Star or the Morning Star, it is brilliant and romantic. Hence the association with Venus the goddess of love.

There is some possibility that the crescent of Venus might be discernible with the unaided eye. In principle the eye can resolve about 15 arc seconds for an 8 millimeter pupil. The diffraction limit for resolving power (rp in arc seconds) for a lens, or pupil of the eye, is $rp = 120/D$, where D is the diameter in millimeters. Persons having exceptional visual acuity may then be able to make out that the planet is not a point source[35]. It should be easy to see

Table 6.14. *Orbits, colors, and brightness of planets.*

Planet (color)	Mean solar distance (AU)	Orbital period (days) sidereal	synodic	Albedo	Magnitude opposition /elongation	near conjunction
Mercury (brown-gray)	0.387	86.96	115.9	0.07	−1.2	+1.1
Venus (white)	0.723	224.68	584.0	0.59	−4.3	−3.3
Earth (bluish)	1.000	365.26		0.35		
Mars (yellow-brown)	1.524	686.95	779.9	0.15	−2.8	+1.6
Jupiter (creamy yellow)	5.203	4332 d (12 y)	398.9	0.44	−2.5	−1.4
Saturn (creamy yellow)	9.539	10,759 d (29 y)	378.1	0.41	−0.4	+0.9

Note: At conjunction the planets become invisible. The last column is meant as a guide to brightness just outside exact conjunction.

Venus against the solar disk during one of its rare transits. The next transits, which tend to run in pairs, will be on June 8, 2004 and June 6, 2012.

Mars orbits the sun at 1.52 AU. Always beyond the earth, it makes a full circle through the sky in its synodic period of 780 days. Mars has a ruddy color and, like the moon, brightens considerably near opposition[36] (see Glossary).

Jupiter and Saturn are both a creamy yellow. If Jupiter were not so bright, four of its natural satellites could be detected visually (magnitudes +5.1 to +6.3). With binoculars these moons can be observed to move from night to night, often being occulted or becoming invisible by being superposed on the planetary disk.

The apparent path of an outer planet across the sky is compounded by the earth's own motion. Superimposed on the west-to-east movement of these planets is a looping trajectory called retrograde motion. This is nothing more than the parallax of the planet arising from the earth's orbital motion. During retrograde the planet moves east-to-west.

6.15 Meteors

On a clear night perhaps ten 'shooting stars' can be seen per hour. They often leave a faintly luminous trail and sometimes break up into several lesser streaks. Heated by atmospheric friction, meteors are tiny bits of rock or iron alloy that strike the earth's atmosphere at between 12 and 75 kilometers per second. Billions dash into the upper atmosphere each day to deposit between 10 and 100 tons of cosmic dust on the earth.

A meteor bright enough to light up the whole sky, or be seen in the daytime, is called a fireball; if it explodes it is a bolide. Such a great meteor may leave a trail of luminous gas lasting for several minutes.

In space they are called meteoroids; if they reach the ground they are meteorites. Like the earth and other planets, meteoroids are part of the debris left over from the formation of the solar system.

Most visible meteors are a few millimeters across or about the size of an apple seed. Despite this minute dimension, a brilliant shell of glowing gas, a meter or so in cross-section, is heated to several thousand degrees along the entry path. Lasting about one second, this blaze accounts for the light we see. Triangulation and radar measurements show that meteors are brightest at an altitude between 130 and 80 kilometers. Those large enough to make it to the ground are slowed by friction and cool until they are no longer visible below 80 kilometers.

There are two kinds of meteors: shower meteors and sporadic meteors. Meteor showers occur at regular intervals each year (Table 6.15), because they are debris from periodic comets. They seem to radiate out from a fixed point in the sky, the radiant, and they are named for the constellation of its location. Shower meteors in reality travel in parallel paths; this fanning out is due to perspective. On November 12, 1833 a swarm of Leonids was encountered which led to an estimated count rate of 200 000 meteors per hour. No shower meteor is known to have reached the ground intact. This is because comets are made primarily from ice and dust grains which are too small to survive their passage through the atmosphere.

Sporadic meteors are simply meteors from the larger solar system debris cloud. They are always more numerous after midnight. The morning sky is on the leading side of the earth in its orbit and that hemisphere runs into more meteors. In the evening sky, the meteors have to catch up with us and the slower ones never make it.

Table 6.15. *Dates for meteor showers.*

Name	Maximum	Interval	Typical hourly rate
Quadrantids	Jan 3	Jan 1–6	60
Lyrids	Apr 23	Apr 20–24	10
Aquarids	May 6	May 1–10	35
Aquarids II	Jul 30	Jun 20–Aug 14	20
Perseids	Aug 12	Jul 23–Aug 20	75
Orionids	Oct 21	Oct 17–27	25
Taurids	Nov 4	Oct 20–Nov 30	10
Leonids	Nov 17	Nov 14–19	6
Geminids	Dec 13	Dec 8–15	75

6.16 Grazing meteors

On August 10, 1972, a large meteor nearly hit the earth[37]. It was bright enough to be seen during the day. Heading north across Utah, Idaho, Montana, and into the Province of Alberta, it took some 30 seconds to pass overhead. Some people caught it on film. The fireball was also observed by an Air Force satellite which meant that its orbit could be computed. Traveling at 34.8 kilometers per second it overtook the earth and crossed our orbit plane from south to north. It entered the atmosphere at a grazing angle and came to within about 60 kilometers of the surface before escaping after a blazing trajectory totaling some 1500 kilometers. What a sight that must have been!

6.17 Comets

To us a comet is a fuzzy spot with a tail, occasionally found in the early evening or pre-dawn heavens. A comet like Halley's is predictable. Others turn up with no warning. Most comets are never seen except by large telescopes, but a few attain naked-eye visibility. The great comet of 1843 had a tail that at one time stretched across a third of the sky. Sometimes there is a filamentary or knotted texture to the tail. On rare occasions a weak 'anti-tail' has been observed pointing towards the sun. Although photography reveals some coloration, comets are too faint to stimulate the color-sensitive cones of the eye.

Comets are chameleons. Their lives are spent as cold invisible aggregates of ice and dust. Possibly only once are they anything else. This is when their orbits bring them close to the sun, they heat up, and the evaporating ice lofts dust off their surface. This dust is then swept back away from the sun by the solar wind and radiation pressure to form a tail. Comet tails point away from the sun regardless of the direction of motion. The exceptions, anti-tails, probably are made of larger particles which show up momentarily when the earth passes through the orbital plane of a comet.

Where do comets come from? Far beyond the realm of planetary orbits is thought to exist a vast volume of space occupied by comets in a holding pattern called the Oort Cloud[38]. Astronomers estimate that some $10^{11} – 10^{13}$ comets occupy this 'parking lot' at 30×10^4 to 100×10^4 AU. (The nearest star, Proxima Centauri, is 271×10^4 AU distant). The pace of comets in the cloud is very slow, perhaps 1 kilometer per hour. For reasons not yet understood a comet can be deflected sunwards. Perhaps there is an interaction between themselves or the gravitational nudge from a passing star.

The arrival of comets from the cloud is unpredictable. Often they are first picked up by comet hunters. In September, 1965, two such amateur astronomers, Kaoru Ikeya and Tsutomu Seki, independently discovered a naked-eye comet in the morning sky[39]. Already it was within 1 AU of the sun and seemingly headed straight for it. By noon on October 20 Ikeya–Seki was at perihelion and easily visible in the daytime sky near the sun (Figures 6.17A, 6.17B). After perihelion its tail became a spectacular night-time feature with a length of 50°. Fading rapidly the comet disappeared to the unaided eye in early November.

Fig. 6.17A Sungrazer Ikeya–Seki on October 21, 1965, when about 480000 km from the sun's surface. (Photo by Tokyo Observatory)

Fig. 6.17B Comet Ikeya–Seki with its sweeping dust tail as seen in the morning sky of October 28, 1965, from Kitt Peak. (Photo by Michael Belton)

Comet arrivals from the cloud are a one time thing. As far as we can tell, their orbital paths are parabolic or nearly so. This means that a comet does not return within a reasonable time, say 100 000 years or more. But if a comet from the cloud passes near to Jupiter it can be diverted into an elliptical orbit and become a 'periodic comet'. Halley's Comet is our brightest periodic comet, having a period of 76.9 years and its passages have been recorded in history since 240 BC. With the exception of Halley's Comet, predictable comets are too faint to be readily seen with the naked eye. Perhaps repeated encounters with the sun have worn them out. In any case the best comets in the future will all be surprises!

Fig. 6.17C Comet Hale–Bopp from Frazer Mt Lookout, California, on April 4, 1997. The whitish tail is scattered light from dust, the bluish one originates from the fluorescence of atoms and molecules.

6.18 Zodiacal light and gegenschein

Well after sundown when twilight has faded, a broad diffuse wedge of light is found in the western sky reaching toward the zenith from the horizon[40]. It lies along the ecliptic and is about as bright as the Milky Way. Moving as though a part of the field of stars, it sets about three hours after the sun only to reappear roughly three hours before sunrise in the east. This is the zodiacal light. The best time of year to see it is in the spring (northern hemisphere) when the ecliptic makes its greatest angle with the horizon (Figure 6.18).

The zodiacal light is sunlight scattered from interplanetary dust. These particles orbit the sun in a flattened pancake-like swarm confined about the same plane as the planets and concentrated in the inner solar system.

On dark nights there is yet another patch of faint light that is not fixed among the stars. Look along the ecliptic to a position you might estimate as the antisolar point. Called the gegenschein (German for 'counter-glow'), it is roughly 6° × 10° in size, being elongated along the ecliptic. All-sky photographs reveal that the gegenschein is also a part of the zodiacal light[41-3]. The enhancement near the antisolar point is simply a consequence of preferential backscattering.

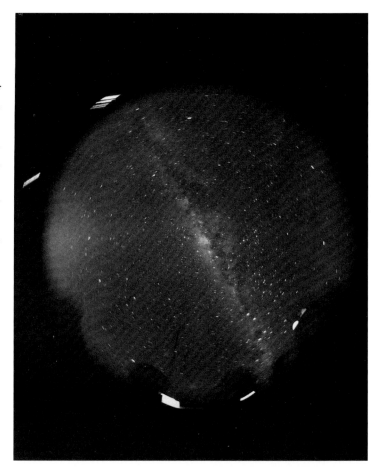

Fig. 6.18 Zodiacal light, to the left, the Milky Way with its gaseous dark rift, and the galactic center directly overhead. The two tiny faint elongated patches at the bottom are the Magellanic Clouds. There is a suggestion of the airglow seen as a brightening of the sky toward the horizon. Photographed at Cerro Tololo in the southern hemisphere by A. Hoag.

6.19 Stars and the stellar magnitude scale

Although the number of stars visible to the unaided eye may seem countless, the actual number is about 6000. And only half of these are visible at one time because we can only see half of the celestial sphere. Under very special conditions[34] the faint limit for stars may be extended, but this quoted total number is correct for general sky-gazing.

The faintest stars visible to the naked eye are about 300 times dimmer than Sirius, the brightest. Their apparent brightness thus covers a wide range. For this reason, a logarithmic scale has been devised that matches the non-linear response of the eye. In 1856 Norman Pogson formalized an ancient and somewhat arbitrary system by assigning the stars Aldebaren and Altair the brightness of 'first magnitude.' Stars 2.5 times fainter (actually 2.512... = fifth root of 100) were termed second magnitude, those dimmer by another factor of 2.5 were third magnitude, and so on.

Like the centigrade temperature scale, the Pogson scale has its zero point displaced from the beginning; objects brighter than the first magnitude have negative magnitudes. For example Sirius, the brightest star, has a magnitude of −1.6. At their brightest, Mars, Jupiter, and Venus are respectively, magnitude −2.8, −2.5, and −4.3. The full moon is −12.5, and the sun blazes at −26.5 magnitudes. The faintest star detectable by eye is about +6.5 magnitudes. The number of stars visible increases rapidly as we go fainter (Table 6.19).

Most stars have surface temperatures between 3000 and 30 000 °C and emit light roughly like blackbody radiators. Their color depends on temperature; 3000–4000 °C red, 4000–5400 °C yellow, 5400–6000 °C white, 6000–10 000 °C bluish-white, and above 11 000 °C bluish. Sirius is blue-white, Capella is yellow, Arcturus is perhaps orange, and Betelgeuse and Antares are decidedly reddish. The sun, with a surface temperature of 5700 °C defines 'white'. Colors of the stars are best revealed by star-trail photographs[44] (Figure 6.19). Our retinal rods are color blind, and only the brightest stars manage to excite the color-sensitive cones. Looking through binoculars while weaving them about helps somewhat to see color.

Table 6.19. *The magnitude scale and star counts.*

Mag. limit	Relative Brightness	No. stars visible
1.0	1.0	5
2.0	0.40	25
3.0	0.16	91
4.0	0.063	530
5.0	0.025	1620
6.0	0.010	4850
7.0	0.004	14 300

Fig. 6.19 Star trails of the setting constellation of Orion over the 2.4 meter telescope at Kitt Peak. Some star colors are evident. The pinkish trail just above the dome belongs to the Orion Nebula. Betelgeuse, a red giant, is the trail ending at the lower right. Notice the extinction and reddening of starlight by the dusty atmosphere, and the fact that the star trails narrow towards the horizon, showing that the camera film image is broadening the star trails by irradiation. But the Orion Nebula, being an extended source, does not narrow in the same way.

6.20 Daytime visibility of stars

Sometimes a star-like object will catch our attention during daylight. Most often it is Venus, but Jupiter is another candidate. Could we see a bright star? In fact Sirius has been sighted 20 minutes before sunset[45], but this indicates exceptional conditions and a superb eyesight. As a rule stars cannot be seen in daytime.

Since the time of Aristotle, the idea has persisted that bright stars are visible during the day provided the skylight is blocked by viewing through a tube or from the bottom of a well. Modern observers[46–8] have tested this concept and found it to be untrue. Indeed, the absence of the skylight by shielding may actually inhibit detection by inducing dark adaptation and thus causing the retina to saturate from the surrounding skylight.

6.21 Constellations and star names

The sky is divided into 88 constellations whose boundaries, as defined by the International Astronomical Union, are irregular and rather arbitrary. Passing through fourteen of the constellations is the band of the Zodiac, roughly the annual path of the sun and planets through the sky (the sun's exact track being the ecliptic). For naked-eye astronomers these constellations serve as a road map to celestial objects. (See *Norton's Star Atlas*[49] or the *Cambridge Star Atlas 2000.0*[50]. To find any constellation, consult the star charts in one of these books.)

Bright stars are designated by a Greek letter approximately ordered by brightness and followed by the genitive form of the constellation in which the star resides. Thus the brightest star in Boötes is α Boötis. A few constellations, such as Ursa Major, have alternative common names (the Big Bear or Big Dipper).

6.22 Double stars, variable stars, novae, and supernovae

Many stars come in pairs and a few of these can be resolved with the naked eye. Such stars provide tests of your visual acuity. The second star from the end of the handle of the Big Dipper, ζ Ursae Majoris, is the well-known double Mizar and Alcor, whose separation is 11 arc minutes but with a two magnitude difference in brightness. Another but more difficult case is ε Lyrae with an angular separation of 3 arc minutes. Most double stars orbit each other and are called binaries; others are close merely by chance.

Some stars vary in brightness with time. One example is Algol (β Persei), an eclipsing binary. Algol was known as the 'demon' star representing the sinister winking eye of the Gorgon Medusa[51]. The earth happens to lie almost in the plane of Algol's orbit so that one star eclipses the other. It undergoes an eclipse every 2 days 21 hours at which time it drops by over a magnitude.

Other stars physically pulsate with accompanying brightness change. Mira (o Ceti) has a period of 11 months and an enormous magnitude range from +2 to +9. This means it goes from being fairly bright to fainter than we can see. It disappears.

Novae are stars that, usually only once, increase in light rapidly to be followed by a gradual fading. Novae are concentrated in the Milky Way since this is the region of greatest star density. Naked-eye novae are infrequent, perhaps one every decade.

A nova starts out as a binary star, one of them being a white dwarf. The two stars are so close together that their mutual gravity distorts the larger one to the extent that it squirts its outer atmosphere onto the white dwarf. When enough material has accumulated, the surface layers of the white dwarf become unstable and undergo a nuclear explosion. Material is then blown into space and it is this event that causes the star to brighten so much.

Supernovae are the biggest guns in the universe. In the course of a few days the output of a typical supernova will rival that of an entire galaxy, or roughly the radiative energy of sun summed over its whole lifetime. This cataclysmic event represents the virtual destruction of a star and is extremely rare; about one supernova is observed per century in our galaxy. In 1987 a supernova appeared in the Large Magellanic Cloud, a neighboring galaxy of the Milky Way. It reached third magnitude and so was easily visible with the naked eye. As vividly described by Verschuur[52] in *Cosmic Catastrophes*, a nearby supernova could spell the end of life on earth. The tremendous accompanying gamma ray shower would likely destroy our protective layer of atmospheric ozone. Yes, even at stellar distances!

6.23 Star clusters and nebulae

Many stars are found in associations or clusters. Best known of these is the Pleiades, or 'seven sisters'. This is in reality an open cluster of hundreds of stars. Of these, six are readily visible, perhaps eight more can be recognized if the sky is dark and your eyes exceptional. The astronomers W. Pickering and L. Jacchia were reported[34] able to see 20 stars! This cluster in Taurus is a gravitationally related group that are traveling together through space. In Cancer, the globular cluster of thousands of stars can be seen as a fuzzy patch on a moonless night.

Some other fuzzy patches are not stars at all but glowing clouds of dust and gas. The brightest of these is the Orion Nebula in the sword of Orion. By star-trail photography it displays a distinctive red color because of hydrogen 656 nanometer emission (see Figure 6.19).

6.24 The Milky Way

This band of faintly glowing stars stretches almost from horizon to horizon. Singly the stars are unresolved and invisible, but taken together they represent the plane of our galaxy. Although seeming much brighter than other regions of the sky, the difference is only a factor of two. The summer Milky Way is more intense than the winter one because in the former we are looking towards the galactic center in Sagittarius where the star concentration is greatest. The sun, whose position lies about mid-way out from center in the plane of the Milky Way, is just one of the 10 000 million stars that make up our 'island universe.' Were we to find ourselves at the galactic center, the nighttime sky would contain such a crowding of stars that its illuminance would equal ten full moons. Sprinkled along the Milky Way are some of the brightest stars in the sky as well as absorbing clouds of dark gas. An example of a dark cloud is the 'coalsack' in the Southern Cross.

6.25 Galaxies

Our galaxy, the Milky Way, has two companion galaxies: the Large and Small Magellanic Clouds at a distance of 160 000 light years. These are southern hemisphere objects (Figure 6.18). Another galaxy is the distant Andromeda Galaxy, which is an easy naked-eye target on a dark night in the northern skies of autumn. It is two million light years away, which means its light that we see tonight left that galaxy long before Homo Sapiens came into being.

6.26 Light of the night sky

If we hold up our hands in front of us, we can determine that the night sky is not indefinitely dark. Light from the night sky has a variety sources: nearby urban glows, airglow and aurora in the terrestrial atmosphere, interplanetary zodiacal light, starlight, and the diffuse galactic light of the Milky Way[53]. Discounting city lights, the relative contribution of these sources to ground illumination is shown in Table 6.26.

Table 6.26. *Sources of light in the night sky.*

Source	Intensity
airglow	16%
aurora	variable
zodiacal light	42%
integrated starlight	36%
galactic light	6%

6.27 Urban glows

From a distance on a clear, dark night, the presence of scattered light from a city creates a localized urban glow. Figure 6.27A is a full-sky picture that at once discloses the presence of Phoenix, Tucson, and even tiny Nogales by their urban glows. Phoenix is visible from 150 kilometers, metropolitan Los Angeles from 250 kilometers. Figure 6.27B is a satellite photograph of urban glows covering the eastern half of the North America.

The detrimental effect of city lights to astronomical observing obviously depends on their proximity. Unfortunately, the contribution to skylight grows with time. Mount Wilson Observatory, located at an elevation of 2000 meters in the San Gabriel Mountains above the Los Angeles basin, has suffered fatally from the growth of city lights. When the valley is covered by advection fog, however, the sky over the mountain almost reverts to its former darkness.

Monochromatic street lighting can add an element of color. Most cities have areas of blue-green mercury lamps which contrast in a quilt-work fashion with the yellow from sodium illumination.

Fig. 6.27A Urban glows photographed from Kitt Peak at the McMath–Pierce Telescope with a 6 millimeter 'fisheye' lens by Dean Kitelson. Counter-clockwise at 2:30 is Tucson, at 4:30 is Nogales, a small town on the Mexican border. At 11:30, behind the telescope mirrors, is Phoenix. Along the western horizon from 8:00 to 11:00 is the twilight arch. Also visible is the zodiacal light and the Milky Way.

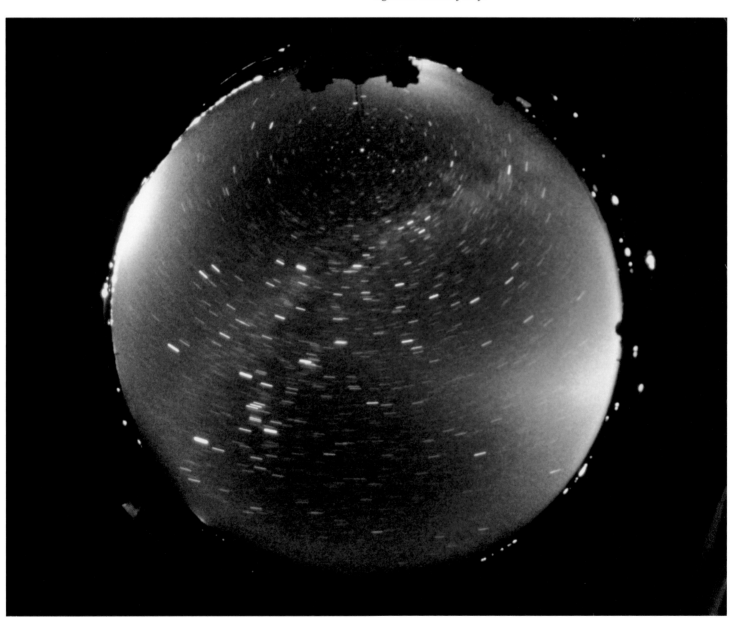

Fig. 6.27B Urban glows in eastern North America as recorded by US Air Force satellite. A faint network interconnects cities and towns.

6.28 Starlight

Astronomers have mapped the sky with large telescopes that detect stars down to +25 magnitude and fainter. In this way they can separate starlight from other light sources such as the airglow and nebulae. They find that if stars were uniformly distributed over the sky a general glow would result equivalent to 51 stars as bright as Sirius (which has a magnitude of −1.58).

Just as dust in the solar system gives rise to zodiacal light, interstellar dust in the galaxy scatters starlight to create an extra 6% illumination called diffuse galactic light. This light probably escapes notice to the unaided eye.

6.29 Olbers' paradox

In 1744 the Swiss astronomer Philippe Cheseaux assembled a powerful argument concerning the stellar night sky. Speculating that the universe was infinite and that there were an infinite number of stars in it, he argued that in any direction our sight-line should eventually strike a star. Taking the sun as a typical star, Cheseaux concluded that the entire the sky should be as bright as the sun. Heinrich Olbers independently came to the same conclusion in the nineteenth century. In as much as the night sky is relatively dark, this puzzling state of affairs became known as Olbers' paradox.

In recent years we have learned that the universe is not infinite and does not contain an unlimited number of stars. But that does not alter Olbers' paradox much. What does matter is that the universe is expanding. Doppler red-shifted light from the most distant parts carries far less energy to the eye. Even more important is the finite age of the universe. Light from the most distant galaxies has simply not yet reached us[54,55]. And that is why it is dark at night.

References

1 Noyes, R.W., 1982, *The Sun, Our Star*, Harvard Univ. Press, Cambridge, Mass.

2 Eddy, J.A., Stephenson, F.R., and Yau, K.K.C., 1989, 'On pre-telescope sunspot records', *Quarterly Journal of the Royal Astronomical Society*, **30**, 65-73.

3 Mossman, J.E., 1989, 'A comprehensive search for sunspots without the aid of a telescope', *Quarterly Journal of the Royal Astronomical Society*, **30**, 59-64.

4 Schaefer, B.E., 1991b, 'Sunspot visibility', *Quarterly Journal of the Royal Astronomical Society*, **32**, 35-44.

5 Keller, H.U. and Friedli, T.K., 1992, 'Visibility limit of naked-eye sunspots', *Quarterly Journal of the Royal Astronomical Society*, **33**, 83–89.

6 Giovanelli, R., 1984, *Secrets of the Sun*, Cambridge University Press, Cambridge.

7 Meeus, J., 1995, *Astronomical Tables of the Sun, Moon, and Planets*, Willmann-Bell, Richmond, Virginia.

8 Meeus, J., 1982, 'The frequency of total and annular solar eclipses for a given place', *Journal of the British Astronomical Association*, **92**, 124.

9 Morrison, L.V., Stephenson, F.R., and Parkinson, J., 1988, 'Diameter of the Sun in AD 1715', *Nature*, **331**, 421.

10 Zirker, J.B., 1984, *Total Eclipses of the Sun*, Van Nostrand Reinhold Co., New York.

11 Littmann, M. and Willcox, K., 1991, *Totality – Eclipses of the Sun*, University of Hawaii Press, Honolulu.

12 Codona, J.L., 1991, 'The enigma of shadow bands', *Sky and Telescope*, **81**, 482.

13 Langley, S.P., 1896, *The New Astronomy*, Houghton, Mifflin, and Co. Boston.

14 Ranyard, A.C., 1879, 'A compilation of total eclipse phenomena', *Memoirs of the Royal Astronomical Society*, **41**, 1–792.

15 Schaefer, B.E., 1991a, 'Eclipse earthshine', *Publications of the Astronomical Society of the Pacific*, **103**, 315–316.

16 Mossley, J., 1991, 'The Sun sets twice on January 4', *Griffith Observer*, **55**, 2–10 (Dec).

17 *The Astronomical Almanac 1991*, US Government Printing Office, Washington.

18 Fiala, A.D., DeYoung, J.A., and Lukac, M.R., 1986, 'Solar eclipses, 1991–2000', *US Naval Observatory Circular No. 170*.

19 King-Hele, D. and Eberst, R., 1986, 'Observing artificial satellites', *Sky and Telescope*, **71**, 457.

20 Miles, H., 1974, *Artificial Satellite Observing and Its Applications*, Faber and Faber, London.

21 Taylor, G.E., 1986, 'Geostationary satellites', *Sky and Telescope*, **71**, 557.

22 Maley, P.D., 1987, 'Specular satellite reflection and the 1985 March 19 optical outburst in Perseus', *Astrophysical Journal*, **317**, L39.

23 MacRobert, A., 1985, 'The Aries Flasher: A status report', *Sky and Telescope*, **70**, 54.

24 Maley, P.D., 1986, 'Photographing Earth satellites', *Sky and Telescope*, **71**, 563.

25 Livingston, W. and Talent, D., 1990, 'Those elusive geostats', *Sky and Telescope*, **80**, 319.

26 Soop, E.M., 1994, *Handbook of Geostationary Orbits*, Kluwer, Dordrecht.

27 Houzeau, J.C. and Lancaster, A., 1889, *General Bibliography of Astronomy to the Year 1880*, Vol 1, Part 1, pp 40–43, in French, reprinted 1964, Dewhirst, D. (ed), The Holland Press, London.

28 Alter, D., Cleminshaw, C.H., and Phillips, J.G., 1963, *Pictorial Astronomy*, Thomas Crowell, New York.

29 Meketa, J.E., 1987, 'Viewing libration with the naked eye', *Sky and Telescope*, **74**, 63 (July).

30 Bobrovnikoff, N.T., 1967, 'Pre-telescopic topography of the Moon', in *Modern Astrophysics*, a Memorial to Otto Struve, Hack, M. (ed), Gordon and Breach, New York.

31 Ashbrook, J., 1967, 1972, 'Astronomical scrapbook' in *Sky and Telescope*, **34**, 92 (Aug 1967), and 43, 95 (Feb 1972).

32 di Cicco, D., 1989, 'Breaking the new-moon record', *Sky and Telescope*, **78**, 322 (Sept).

33 O'Meara, S.J., 1992, 'Strange lunar eclipses', *Sky and Telescope*, **84**, 687–689 (Dec).

34 Bobrovnikoff, N.T., 1978, *Astronomy before the Telescope, Volume I, The Earth–Moon System*, Pachart Publishing House, Tucson.

35 Hostetter, C., 1988, 'Venus' crescent', *Sky and Telescope*, **75**, 461 (May).

36 Young, A.T., 1985, 'What color is the solar system?', *Sky and Telescope*, **69**, 399.

37 Jacchia, L.G., 1974, 'A Meteorite that missed the Earth', *Sky and Telescope*, **48**, 4–9 (Jul).

38 Brandt, J.C. and Chapman, R.D., 1981, *Introduction to Comets*, Cambridge University Press, Cambridge.

39 Marsden, B.G., 1965, 'The great comet of 1965', *Sky and Telescope*, **30**, 332.

40 Blackwell, D.E., 1960, 'The zodiacal light', *Scientific American*, **203**, 54 (Jul).

41 Roosen, R.G., 1970, 'The gegenschein and interplanetary dust outside the Earth's orbit', *Icarus*, **13**, 184.

42 Solberg Jr., H.G. and Minton, R.B. 1966, 'Photographing the gegenschein', *Sky and Telescope*, **31**, 380.

43 Struve, O., 1951, 'Photography of the counterglow', *Sky and Telescope*, **10**, 215.

44 Murdin, P., 1987, 'Colours of the stars', *Journal of the British Astronomical Association*, **97**, 204.

45 Henshaw, C., 1984, 'On the visibility of Sirius in daylight', *Journal of the British Astronomical Society*, **94**, 221.

46 Weaver, H.F., 1947, 'The visibility of stars without optical aid', *Publications of the Astronomical Society of the Pacific*, **59**, 232.

47 Smith, A.G., 1955, 'Daylight visibility of stars from a long shaft', *Journal of the Optical Society America*, **45**, 482.

48 Hughes, D.W., 1983, 'On seeing stars (especially up chimneys)', *Quarterly Journal of the Royal Astronomical Society*, **24**, 246–257.

49 Norton, A.P. and Inglis, J.G., 1980, *Norton's Star Atlas*, Sky Publishing Co., Boston.

50 Tirion, W., 1991, *Cambridge Star Atlas 2000.0*, Cambridge University Press, Cambridge.

51 Room, A., 1988, *Dictionary of Astronomical Names*, Routledge, London.

52 Verschuur, G.L., 1978, *Cosmic Catastrophes*, Addison-Wesley, Reading, Mass.

53 Roach, F.E. and Gordon, J.L., 1973, *The Light of the Night Sky*, D. Reidel Pub. Co., Dordrecht-Holland.

54 Harrison, E., 1987, *Darkness at Night: A Riddle of the Universe*, Harvard University Press, Cambridge, Mass.

55 Wesson, P.S., Valle, K., and Stabell, R., 1987, 'The extragalactic background light and a definitive resolution of Olbers' paradox', *Astrophysical Journal*, **317**, 601.

7 Observing

We strolled along the Redondo Beach Pier with astronomer Dale Vrabec, enjoying the warm summer evening. Pausing to gaze at the moonlight on the water, Dale exclaimed
'Hey. Look at all those circles.'
'Where?'
'There, in the moon's reflection, all those little circles, coming and going.'
'What?'
After some patient coaching by Dale, we suddenly saw them. They were all over the place. In fact, once we got the hang of it, they were absolutely obvious! Was it possible that no one before, in all the millennia that people have stared at the moon's glitter, had noticed these remarkable little circles?

A recollection from our 'discovery' of moon circles in May 1975

This chapter is about basic 'tools of the trade' and observing technique. How the eye behaves – and misbehaves. Given patience and enough time, nature will parade most optical phenomena past our eyes right at home. The majority of the photographs in this book were not planned, but were taken on the spur of the moment in the course of our daily routines. While many phenomena can be readily seen with the unaided eye, a number of techniques can be used to see them more often and perhaps better. Most important, each time you go outside, take a moment and scan the sky and landscape. And always have a loaded camera handy.

All of us tend to see the expected and are blind to the unaccustomed. R.D. Laing put it this way: 'The range of what we think and do is limited by what we fail to notice'. In a similar vein Louis Pasteur said 'Chance favors the prepared mind'.

Information is one antidote. Be aware that the glitter from lunar reflections on water forms closed figures, that there is more to a rainbow than the primary arc, that ripple-like patterns are likely found inside the arc, that there is a secondary bow which is inverted as to color, that up to four arcs sometimes appear when proximate to water, that a polaroid can completely extinguish parts of a rainbow, and so on. All these events become things we can look for and then finally perceive – once we know about them.

HUMAN VISION

7.1 The eye

Your eyes are the most important pieces of observing equipment you have. By any standard they are amazing devices[1]. At most visible wavelengths they exceed the sensitivity of photographic film and can discern color differences in the green as small as 1 nanometer[2]. Within scenes ranging in brightness B from 1 to 100 000 candles/per square meter, the minimum change of brightness $\Delta B/B$ we can detect is about 2%. Owing to our eyes' approximate logarithmic response[3], we are able to navigate a path by either sunlight or moonlight, a difference of a million in illuminance. Each eye has a field of view of over a 100° and together our eyes are able to sense the presence of objects over an entire hemisphere, especially if they move.

In several ways the eye resembles a camera. Light enters through a transparent outer membrane called the cornea and passes through the pupil which defines the eye's aperture. Depending on light level, the diameter of the pupil varies between 2 and 8 millimeters. Exceptions are the elderly, whose maximum opening is generally no greater than 5 millimeters for those over 50 years[4, 5]. The main purpose of the iris is to stop down the aperture to achieve maximum acuity at high light levels. Next is the crystalline lens which, in combination with the cornea, focuses light on the retina. The effective focal length is about 17 mm. This is the sensitive part of the eye analogous to film in a camera. Although our field of view covers a wide angular area, there is a central region, the fovea, which is more richly endowed with photoreceptors than any other part. When looking directly at an object, the feature of interest is directed to the fovea. While looking at any word on this page attempt to read words only a centimeter or so away; it cannot be done. In fact return to that semicolon and try to scrutinize both parts simultaneously. The angle between them is only 0.3°. The remaining field of view is peripheral and falls on retinal surroundings, the macula.

Each eye has a blind spot at the connection to the optic nerve. We do not perceive this spot in our visual field. Instead the surrounding field fills it in, a process not fully understood[6]. Thanks probably to early learning, several other potential distractions of the eye are suppressed: that nose protruding into the scene; an image which is sharp only at the fovea and becomes indistinct toward the periphery; the eye's need for constant movement (Section 7.5).

Sensor elements on the retina are of two kinds: cones for high light level viewing (photopic vision), and rods for low light conditions (scotopic vision). There are approximately five million cones, a hundred million rods, but only a million nerve connections to the brain. This means there are 5–100 sensors per neuron.

7.2 Photopic (cone) vision

Cones provide our highest angular acuity and they also enable us to see color. These sensors are concentrated at the fovea although they extend more sparsely across the entire retina (excepting at the blind spot). The spectral response of cones has a peak at 560 nanometers (yellow-green) and a nominal range from 400 nanometers to 760 nanometers (Figure 7.2A).

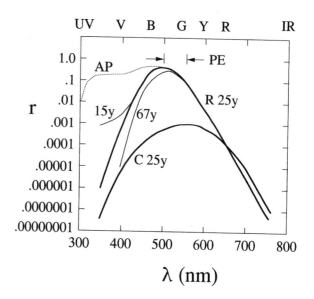

Fig. 7.2A Relative response (r) of the eye for the photopic (high light level) cones (C) and scotopic (dark adapted) rods (R) as it depends on wavelength (λ). Scotopic sensitivity in the blue-violet weakens with age and the applicable age in years (y) is shown. Removal of the cornea results in aphakic vision (AP). The Purkinje effect (PE) takes place in going from cone to rod vision. UV=ultraviolet, V=violet, B=blue, G=green, Y=yellow, R=red, IR=infrared.

Limits of the eye's overall range of sensitivity extends from about 310 to 1050 nanometers, but strong illumination is necessary for sensation at these wavelength extremes. With age our blue and violet response fades as the lens yellows[2]. Although the retina can easily detect high intensity ultraviolet light below 400 nanometers, the lens cannot focus it. This explains why ultraviolet lights always look a bit fuzzy. Removal of the lens, as in a cataract operation (the aphakic eye), results in enhanced ultraviolet sensitivity (Figure 7.2B). In very dim illumination, say by light of the crescent moon, the cones become completely inoperative and we lose all color perception, explaining why the lunar rainbow is white (Section 4.2).

It is commonly believed that the peak of the eye's sensitivity near 550 nanometers coincides with the peak of the solar spectrum. The reasoning is that evolution has optimized the eye's sensitivity to take advantage of the available sunlight. This is wrong. The solar spectrum appears to peak near 550 nanometers only when plotted in wavelength units. When plotted in frequency units it peaks near 880 nanometers. The reason for the confusion is the result of comparing two different kinds of functions. The peaks in sensitivity curves do not change when they are transformed from wavelength to frequency units, but peaks in distribution functions like the solar spectrum or the Planck function do [7,8].

Not only is color vision complicated and not yet understood, the production of color, i.e. the physical mechanisms that cause it, are wide ranging. The reader should consult a book devoted to this subject[9].

Fig. 7.2B Aphakic vision through an ultraviolet transmitting filter reveals a landscape something like the left hand scenes compared to the visible red on the right. In the ultraviolet the sky is white, airlight strong, vegetation darkens, and glass is opaque. The latter fact means that a city will be completely dark at night since most illumination is behind glass.

7.3 Scotopic (rod) vision

On entering a darkened room from daylight conditions, the rods slowly begin to take over as receptors. Something like 20 minutes to an hour is required for them to sensitize fully and achieve dark adaptation. During this interval a chemical substance called rhodopsin is activated on the rods to impart extra response. Rhodopsin absorbs best at 500 nanometers and this becomes the peak in spectral sensitivity for scotopic vision. The result is a wavelength shift to the blue compared to the 560 nanometers peak for photopic vision. This shift is called the Purkinje Effect (Section 7.6).

To call on the rods purposely we may exercise 'averted vision'. This is a technique in which we become aware of the presence of a faint object away from the center of our field-of-view that we cannot see when it is at the fovea.

At 500 nanometers approximately 70 photons are sufficient to elicit a response. Of these only one need be absorbed by each of five rods[10]. If we think about it, there is a flickering or 'noisy' quality to a low level light scene that indicates single neuron activity. This works best when we wake up in the middle of the night in a darkened room[11].

LANDSCAPE IN THE ULTRAVIOLET

Although the eye's retina is sensitive in the UV, its lens is practically opaque there. However, following a cataract operation the eye becomes responsive down to atmospheric cutoff at 300 nanometers. If you have had such an operation, here are a few interesting experiments to perform out-of-doors in the landscape (there is little UV indoors because ordinary glass on light fixtures and windows are opaque shortward of 380 nanometers). UV transmitting filters may be obtained from Kodak or Edmund Scientific (useful are Wratten 18A, Hoya U-340, and Schott UG-1).

The first thing the UV observer notices is that people look strange; fair complexions darken dramatically. The sky is abnormally bright and the sun's disk itself less important. A consequence of the latter is that shadows are indistinct. Distant objects fade to be replaced by the intervening airlight. The interiors of cars and buildings are invisible, as are the eyes of anyone wearing glasses. At night cities are strangely dark.

Ultraviolet photography can substitute for the aphakic eye. Special UV cameras are available but they are expensive. Instead purchase a simple quartz lens and fit it to your camera. Edmund Scientific sells fused silica lenses. Stopped down, say to f/16, such a lens produces excellent results for outdoor static scenes where exposure time may be relatively long. Most color film is purposely insensitive to the UV (a layer rejects the UV to improve sky contrast), so use black and white. Figure 7.2B is an example of such photography.

7.4 Angular resolution of the eye

The smallest separation that two objects can have and still be discerned as separate depends on their brightness and contrast. For photopic vision, the diffraction limit for a 2 millimeter pupil opening is about 1 arc minute at the fovea. For scotopic vision acuity diminishes markedly and is hard to measure, but becomes a fraction of a degree or so.

Just because the eye's resolution is limited to 1 arc minute does not mean smaller bright objects go unnoticed. It is a matter of contrast. We see stars, of course, and even the largest of those is subarc second in reality. Sunspots less than 1 arc minute in diameter are detectable as dark markings on the solar disk under photopic conditions.

7.5 Time constants of the eye

The response time of the eye depends on light level: 20 milliseconds to 1/10 second for photopic conditions; 1 second and longer for scotopic. Pupil size adjustment takes a few seconds. Time constants of up to an hour are associated with the chemical changes that accompany dark adaptation.

Unbeknownst to us, image fluctuations are necessary to stimulate a signal. Experiments show that if the eye's movement is purposely defeated, perception slowly fades until nothing but a gray background remains[12].

THE CASE OF THE DISAPPEARING RAINBOW

When next you see a nice rainbow try the following experiment which demonstrates the need for image movement in vision. First put aside your camera; it cannot record what you are about to perceive. Now fixate on some part of the bow. This is not easy and requires extreme concentration and discipline because the eye naturally wanders. It may help to fixate on a foreground object near the bow, such as a tree branch or power pole. Anyway persist. Suddenly you will become aware that the rainbow has faded away; it has simply vanished! Any motion of your eyes and it will instantly reappear.

What you have just experienced demonstrates that the eye is an AC detector that requires constant movement to function, see Section 7.5. In fact there is nothing peculiar to the rainbow as a test object, except its smoothness and lack of detail makes it an easy target.

7.6 Color vision

The eye is a trichromatic receptor. Our color sensors, the retinal cones, contain three kinds of pigments, call them red, green, and blue, that selectively absorb light (Figure 7.6). Light induced neural stimulations by these cones are sent to the brain for interpretation. Notice that the red and green cones are not well separated in wavelength and that the blue absorption extends well into the ultraviolet where there is little natural light. Taken together, this arrangement yields color vision just as the dye layers in color film produce color pictures.

Certain pairs of colors can add to produce white light as perceived by the eye. These pairs are called complementary colors: yellow and blue-violet; red and blue-green, for instance.

Complementary colors are those that when added produce a colorless sum. The complementary color of a given color can be perceived in its afterimage.

Chromatic adaptation is another effect that may explain 'green shadows' (Section 1.3) made famous by Goethe[13]. Look through a red filter for a minute or so, then remove it and notice that the scene takes on a blue-green hue. Your red response has fatigued to favor the green.

During twilight we encounter two quirks of color vision: color constancy and the Purkinje effect. Color constancy causes an object to retain its accustomed color regardless of illumination conditions. My blue sweater remains blue to me even in the red rays of a low sun, while strangers might assert that it is a dark brown.

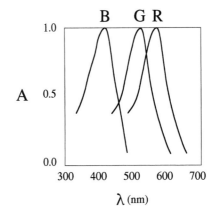

Fig. 7.6 The retinal surface has been examined to measure how the light absorption (*A*) of the cones depends on wavelength λ independent of perception by the brain.

Well after the sun has set our peak color sensitivity shifts from 560 nanometers to 500 nanometers (the Purkinje effect), so that red objects grow abnormally dark and green foliage assumes greater prominence in the landscape.

Not completely understood, color vision has many aspects that are subjective and involve the translation of visual input by the brain. These kind of experiences, which are perfectly real and repeatable, have not been adequately dealt with by any one theory[14–16].

SUBTLETIES OF VISION

7.7 Mach bands

The eye is not to be trusted in the judgement of luminance adjacent to a step change in brightness. Our eye will think it sees non-existent light and dark bands parallel to a contrast step. These regions are called Mach bands (see Figure 2.5B).

Mach bands are present because cones do not operate independently. One cone 'knows' what its neighbor is seeing and responds both to the amount of light falling on it and the adjacent receptors. At a high contrast bright boundary, the interplay between adjacent cones causes a bright stripe where the image becomes dark, and a dark stripe where the image goes bright[17, 18].

7.8 Irradiation

Brightly illuminated areas having sharp boundaries appear to the eye larger than they should. This is called irradiation and examples are found in the apparently exaggerated width of a crescent moon against its dark side, or the notch in the horizon produced by a setting sun (Figure 7.8).

Irradiation is a consequence of scattered light in the receptor and is common to both photographic film and the eye. All the light entering a photographic emulsion is not absorbed; some is scattered within the gelatin layer. When scattered into a region where little or no direct light falls, it exposes the film and the result is indistinguishable from the adjacent bright image. At the high contrast boundary of an overexposed image of the sun, this scattered light spreads across the boundary to produce an enlarged disk. Irradiation in the eye, called the entopic halo, is what we see around street lamps at night.

Fig. 7.8 Setting sun over the Sea of Cortez carves a notch out of the horizon by irradiation.

7.9 After-images

If you gaze intently at a bright scene for several seconds and then close your eyes, or glance at a uniform and darker surface, a strange-colored duplicate of the original is perceived. This is an after-image[2]. Color of the after-image will be complementary to the original. Such after-images can complicate the observation of the green flash or colored shadows, phenomena which are real and not physiological in origin. In our eagerness to see a green flash we may stare too long at the solar disk and the event we seek may be overridden by after-images. Subtle colors in shadows around us may also be lost.

Retinal fatigue, or saturation, is the explanation of after-images. Gazing at a green object leads to a depression of green sensitivity; on turning to a less bright white scene the red and blue cones dominate so that the object area is now sensed as red-blue or magenta.

7.10 Illusions

Here is one we have all experienced: The rising full moon appears huge on the horizon, although an hour later, when well up, it seems normal in size. This is called the moon illusion. Aristotle, Ptolemy, and Confucius all commented on it. At present there exists no consensus regarding the cause of the moon illusion and more than a dozen solutions to the puzzle have been proposed[19].

The so-called size–distance paradox[19] makes sense to us as an explanation. Every object seen in the sky within the atmosphere, for example birds, trees, aircraft, and clouds, appears larger and closer when overhead than near the distant horizon. This is natural because overhead they *are* closer. In this way we come to expect objects to be smaller at the horizon. But for the extra-terrestrial sun and moon such is not the case; they are effectively identical on the horizon as near the zenith. We then interpret this exception as an increase in size at the horizon[20, 21].

In fact, the horizontal lunar disk is diminished 20% in the vertical direction by atmospheric dispersion and 2% in both planes because it is more distant by an amount equal to the earth's radius. So the moon ought to be smaller on rising, not larger. That the moon's size is reasonably constant as it rises can be demonstrated by a time exposure or series of exposures (Figure 7.10A).

Other atmospheric illusions include the 'contrast triangle' (see Glossary), the Mach effect (Section 7.7), and the 'searchlight effect' (Glossary).

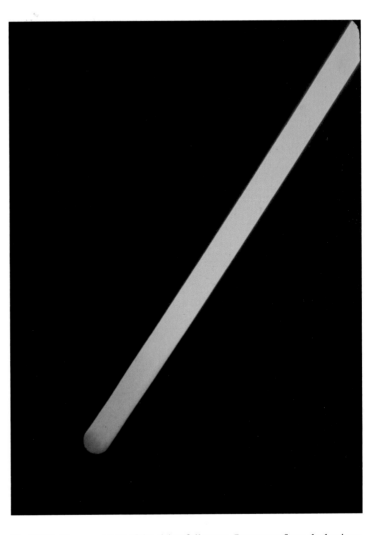

Fig. 7.10A Time exposure of the rising full moon. Departure from the horizon is noticeable in the loss of reddening. Counter to the moon illusion, we find no size change in this photograph.

"*Maynard, I do think that just this once you should come out and see the moon!*"

Fig. 7.10B The moon illusion. (C. Adams, courtesy *The New Yorker*)

7.11 Haidinger's brush

Although the eye is sensitive to polarized light it normally cannot tell us whether or not it is polarized. An exception occurs when the eye is presented with polarization in a featureless background like the clear sky: a faint blue and yellow pattern more than 5° across can be perceived that is called Haidinger's brush (Figure 7.11).

That the brush extends over 5° and has color implies that its origin is in the cones of the macular area of the retina (and not the fovea). Clinical experiments show that the color response of the macula depends slightly on the polarization state (dichroism). Also the outer layer of the cones is birefringent. Although undoubtedly related to Haidinger's brush, neither of these facts directly leads to the observed color pattern and the phenomenon remains something of a mystery[22, 23].

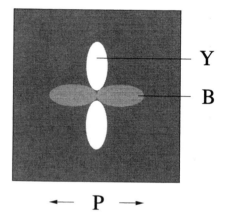

Fig. 7.11 A schematic Haidinger's brush as seen through a polaroid, transmission axis (P), while looking at a white sheet of paper. The actual pattern is diffuse with yellow (Y) and blue (B) components.

THE UNAIDED EYE AS A SENSOR OF POLARIZATION

Under certain circumstances, such as a uniform scene, the eye can analyze polarized light. But to do so requires practice. Out-of-doors find a white automobile, or place a piece of white cardboard in the sun. Take a sheet of linear polaroid, or zero-power polaroid glasses, and view the white surface while rotating the polaroid back and forth. You are looking for Haidinger's brush, a subtle cross in which one arm is bluish, the other yellowish (Figure 7.11). This pattern will rotate with the polaroid. Some people take a while to see it, others recognize it immediately. Once seen, most declare it as obvious.

Now turn your attention to the blue sky which has a maximum polarization about 90° from the sun (Section 2.4). Without a polaroid, look for the brush pattern there, particularly the yellow component. Rocking your head may help. If the brush eludes you, return to looking through the polaroid at the white surface. Eventually almost everyone can detect blue sky polarization naturally.

7.12 Floaters

Have you ever been absently staring out a window and seen ghostly, indistinct little bits of 'something' drifting lazily across your field of view? And if you try to look directly at them, they insistently jump ahead of your gaze? What we see are called floaters (*muscae volitantes* or 'flying gnats'), and not to worry, everybody has them[24].

Floaters are out-of-focus images of detached retinal cells that float around in the converging beam of light from our eye's lens (Figure 7.12). In everyday life they are seldom noticed because their contrast is too low: their angular size is small and they are different in each eye. But against a featureless background they are easily visible. If you look through a pinhole placed close to the eye they become more evident. This improved clarity is a consequence of the increased depth-of-focus that pinholes afford.

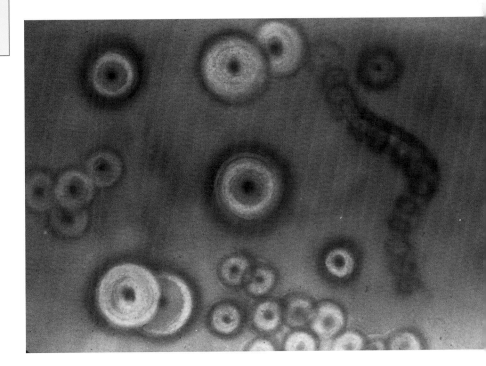

Fig. 7.12 An artist's drawing of floaters in the eye. (By M. Shimamoto)

OBSERVING TOOLS

7.13 Cameras

Ordinary 35 millimeter cameras have been used for most of the photography in this book. We prefer a manual camera, not automatic, in order to ensure that we have full control over the picture taking process. Often there is a need for exposure times outside the normal range, or to be out-of-focus on purpose. A 'bulb' or time exposure option is essential for low light level scenes. Telephoto, or long focus, lenses are helpful in the study of mirages; wide-angle lenses allow us to record entire halo complexes. To photograph single dewdrops and other tiny objects, a close-up lens may be attached to the normal lens.

The majority of our pictures were taken as positive slides (Kodachrome 25 or 64; Ektachrome 64 or 100S). Negative films possess better latitude (exposure range) than positive films but this advantage may be lost in the printing.

Films are color corrected for 'daylight' and artificial or 'tungsten' illumination. The spectral response of color films would extend from 300 nanometers in the ultraviolet to about 700 nanometers in the red, except that daylight type films are overcoated with a layer which is opaque below 400 nanometers. This is done to reduce the loss of contrast from airlight. The tungsten variety film is free of this protection and is therefore suitable for ultraviolet photography. The gamma, or contrast, of both black and white and color films rapidly declines shortwards of 380 nanometers. Ultraviolet light does not penetrate very far into the emulsion layer and the density produced is thereby limited.

7.14 Polarizers

Unpolarized light may be converted to polarized, and polarized light may be analyzed as to plane of vibration (Figure 2.4A) by using a polaroid. Polaroid is available as a camera accessory or as a commercial sheet polarizer material. How completely it polarizes depends on the wavelength being observed. Polaroids should not be used habitually, only for special purposes, because they selectively modify the brightness and may make a picture difficult to understand.

Although not readily available, a 'Savart plate' can be a valuable aid to the visual detection of polarization in broad areas of uniform illumination[25]. The sky viewed through such a device is found to be crossed by regular bands whose strength depends on the degree of polarization (Figure 7.14B). By scanning the sky with a Savart plate one can readily make out the neutral points. Partial polarizations as small as 2% are detectable in this way.

Fig. 7.14A Train window photographed through a polaroid shows stress patterns. Even without a polaroid the unaided eye can sense subtle patterns like these because light reflected obliquely off windows creates a partially polarized illumination environment.

Fig. 7.14B The landscape on Tenerife Observatory, Canary Islands, through a Savart plate. Fringe-like modulation of light is found in the presence of polarization.

Instructions for the construction of a Savart plate are given in books on optical fabrication, for example by van Heel starting at page 401 in Strong[26], or page 74 in Malacara[27].

Riding in a railway car, or other public transportation, we may become aware of subtle image patterns reflected from windows (Figure 7.14A). Safety glass contains plastic material that will retard polarized light in uneven ways. Incident light becomes partially polarized by oblique reflections and this light is further modified, or retarded, to produce these curious slightly colored blotches, Figures 7.14B and also 7.18.

OBSERVING TECHNIQUE

7.15 Measurement of angles

The observer is seldom concerned with linear distances and dimensions. We do not think about the moon as 3500 kilometers across and 400 000 kilometers distant. Rather we refer to the moon as about 0.5° in diameter and at infinity. If we wish to specify its location in the sky, the altitude–azimuth coordinate system tells us its angular position with respect to the north point. Or we can use the equatorial system to fix the moon and planets relative to stars or constellations. Table 7.15 is a guide to gauging angles without instruments.

Table 7.15. *Estimating angles.*

Reference	Angle
complete circle around the horizon	360°
horizon-to-horizon through zenith	180°
horizon-to-zenith	90°
primary rainbow	84° in diameter
22° halo	44° in diameter
35 mm camera's field of view (55 mm lens)	34° × 23°
35 mm camera's field of view (8 mm lens)	132° × 112°
thumb to little finger (fingers spread)	20° (at arm's length)
width of fist	10° (at arm's length)
field of view of 7 × 50 binoculars	7°
length of Orion's Belt	3°
width of finger	1° (at arm's length)
diameter of sun or moon	0.5°

7.16 Universal Time

When an observation may be of widespread interest the time is always given in UT or *Universal Time*. UT is the zone time for 0° longitude which passes through Greenwich, England. The globe is divided into 24 time zones, each nominally 15° in width. For convenience local time zone boundaries are adjusted to fit political boundaries. In addition there may be seasonal adjustments, like *Daylight Saving*, and a few countries insist on being different by half an hour (India, for example). Any world atlas will have a map showing the time zones, or see the *Observer's Handbook*[28].

Tucson, Arizona, lies in the Mountain Standard Time (MST) zone which is 7 hours earlier than UT. If I were to observe a daylight meteor this afternoon at 2:30 MST, I would first say the observation was at 14:30 MST, putting it on the 24 hour system, and then add 7 hours to find it was 21:30 UT. Any other observer worldwide could then readily convert this UT to their own local zone time.

As another example of the use of UT, suppose we want to know if the lunar eclipse of September 26, 1996 is visible in Tucson. Table 6.13A informs us that mid-eclipse takes place at 02:55 UT. Subtracting 7 hours gives us 19:55 MST or 7:55 PM in watch notation. As this is in the evening, the eclipse can be observed.

7.17 Out-of-focus viewing

This is a simple means by which the perceptual door is opened to a class of small scale phenomena which are otherwise invisible. For example, consider the scattering of light from dew drops on a lawn. By placing your eye far closer to a drop than you can possibly focus, and moving to such a position that the glint from reflected sunlight is caught, a disk of light is seen which is crossed by a network of dark lines. In this manner diffraction patterns or the refracted spectrum from a droplet is revealed. On the microscopic scale of a single drop we are seeing the origins of the supernumeraries or the rainbow itself.

The method is also useful in photography. Sharply focused images of snow sparkles are so bright that they saturate the film to appear white, even though they have color. When out-of-focus, the colored patterns are spread out over a larger area, and are less apt to saturate the film or eye (Figures 7.17A, 7.17B, 7.17C). Even ordinary dust particles become attractive when observed in this way (Figure 7.17D).

What happens with out-of-focus viewing? Light scattered from a point-like object has different colors and brightnesses as a function of direction (see Figure 5.1B). When we place our eye (or the camera lens) close to the shining drop we intercept a wide range of angles. In this way we see not one color but a span of color, not one intensity but a whole diffraction pattern (Figure 7.17E). An additional factor, mentioned above, is intensity. In focus the light may saturate the eye or film.

Fig. 7.17A Dew drops on a mimosa leaf viewed out-of-focus: (a) rainbow fragment, (b) diffraction patterns, (c) specular reflection, (d) heiligenschein.

Fig. 7.17B Examined out-of-focus the thread spun by a spider creates color diffraction patterns.

HOW TO PHOTOGRAPH SPIDER WEBS

There are a few tricks to the photography of spider webs in a way which reveals their colors best. First, be aware that spiders only flourish in the warmth of summer, so forget about winter operations. Second, one needs still air (no wind), a dark background and direct sunlight for illumination. These conditions are difficult to realize in the field. Therefore we made a simple loop device, illustrated to the right, to capture webs and convey them to a suitable protected place. A little rubber cement on the loop makes the webs stick.

Your camera should have either a macro lens or an extender so that you can get down to 1:1. A reflex viewfinder is necessary so that you can see the colors you wish to record and determine the degree out-of-focus that is optimum as in Figure 7.17C.

Our experience is that the orb web of the common Cross or garden spider (*Araneus diadematus*) is as good as any, and is widely distributed over the world.

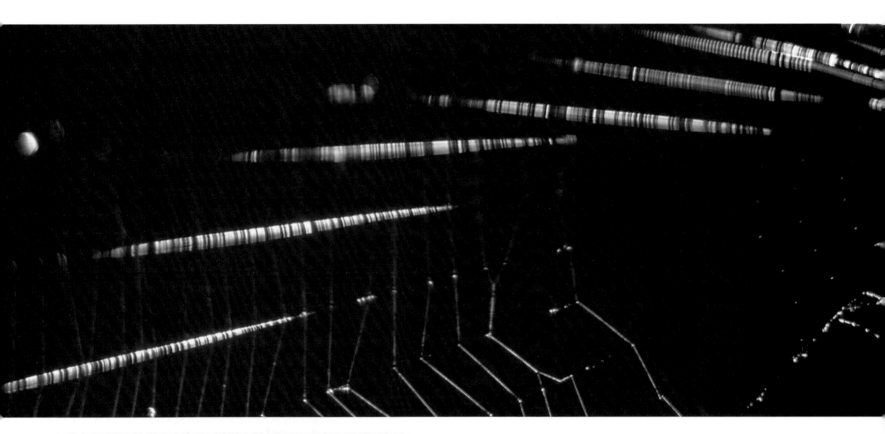

Fig. 7.17C More spider web patterns. Attempts have been made to explain such patterns[1] but we remain puzzled by the detail displayed.

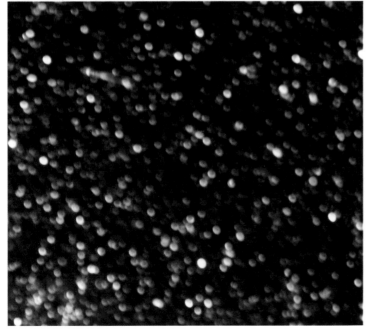

Fig. 7.17D Dust particles take on diffraction colors at different sight angles when viewed out-of-focus. This dust lies on a glass plate, in full sun, but against a dark background. Particle shapes are defined by the camera iris, elongated by a slight movement during exposure, degree of being out-of-focus, and possibly other factors.

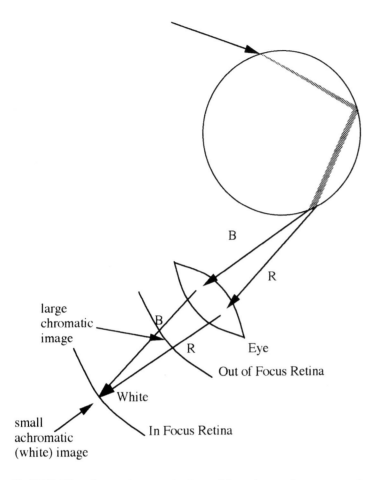

The image contains labels: B, R, large chromatic image, B, R, Eye, Out of Focus Retina, White, In Focus Retina, small achromatic (white) image

Fig. 7.17E When the eye views a water drop rainbow close up the spectrum of the bow spans an area on the retina and one sees the colors. The same applies to out-of-focus images on film. But if we can focus on the drop only a white point is seen.

7.18 Observations from an airplane

Always request a window seat. Get one forward of the wing where you can see both above and below the horizon and where below-horizon scenes are not distorted by jet engine exhaust. Either side of the plane has its advantage: the sunny side for halos, the green flash and flattened suns; the shady side for glories, rainbows, opposition effects, contrail shadows, and the earth shadow.

During takeoff and landing look for the shadow of your plane, especially its three-dimensional form when in hazy air. Notice how quickly the horizon over water blurs with elevation. While flying over bays and seashores, study the water's color and how it depends on depth and bottom conditions. In the tropics, shallow shoals of white sand lead to turquoise water hues. If climbing through thin clouds look for the Spectre of Brocken and glory. At higher elevations, while winding among cumulus canyons, the glory and airplane shadow may repeatedly approach or recede as we encounter clouds containing different droplet sizes. Once above the cloud deck, and provided the angle is right, try to locate the subtle 'cloud contrast bow'.

At cruising altitude be alert for the shadow of your plane's contrail on the cloud deck below. If the sun is low and directly aft, the contrail may produce flickering shadows on the wings. Acoustic shocks can be made visible on the wings as a narrow light and dark strip extending outward along the top of the wing. When flying through cirrus, halos such as subsuns and parhelia can light up unexpectedly. On clear sunny days over land, the opposition effect is ubiquitous.

Sun-glints reveal the presence of wet streets, rivers and ponds by their brilliant specular reflections (Figure 7.18A). The glitter-patch from a distant sea surface is often made golden by the intervening atmosphere to stand out sharply against the whiteness of nearby clouds. Light that has been polarized by oblique reflection off a calm sea is 'retarded' by stresses in your plastic windows to produce interference colors through a polaroid (Figure 7.18B).

At night, on the North Atlantic polar flights, keep an eye out for the aurora borealis. Observe the twilight sequence as you approach Europe and compare this panoramic scene at 10 000 meters with the more familiar view from the ground. If seated opposite from sunrise, study the earth's shadow and accompanying belts of Venus. At high latitudes the earth shadow can persist for an hour[29, 30].

Fig. 7.18A From an aircraft, passing sun-glints map out the location of ponds and other standing water.

Fig. 7.18B Look through an aircraft window with a polaroid and notice how the color of sunlight reflected off the ocean changes (but not that from clouds) according to angle of the polaroid. Manufacturing stresses within the plastic window induce index of refraction variations which, in turn, depend on the state of polarization. Vertically polarized light will travel through the window at a different velocity than that horizontally polarized, an effect called retardance. As the light waves get out of phase we see interference colors.

References

1 Gregory, R.L., 1973, *Eye and Brain*, McGraw-Hill, New York.

2 Le Grand, Y., 1957, *Light, Colour, and Vision*, Chapman and Hall, London.

3 Young, A.T., 1990, 'How we perceive star brightness', *Sky and Telescope*, **79**, 311.

4 Kadlecová, V., Peleška, M. and Vaško, A., 1958, 'Dependence on age of the diameter of the pupil in the dark', *Nature*, **182**, 1520.

5 MacRobert, A.M., 1992, 'A pupil primer', *Sky and Telescope*, **83**, 502.

6 Ramachandran, V.S. and Gregory, R.L., 1991, *Nature*, **350**, 699.

7 Lynch, D.K. and Soffer, B.H., 1999, 'On the solar spectrum and the color sensitivity of the eye', *Optics and Photonics News*, **10**, 28.

8 Soffer, B.H. and Lynch, D.K., 1999, 'Some paradoxes, errors and resolutions concerning the spectral optimization of human vision', *American Journal of Physics*, **67**, 946.

9 Nassau, K., 2001, *Physics and Chemistry of Color*, John Wiley, New York.

10 Hecht, S., Shlaer, S., and Pirenne, M.H., 1942, 'Energy, quanta, and vision', *Journal of General Physiology*, **25**, 819.

11 De Vries, H., 1943, 'The quantum character of light and its bearing on threshold vision, differential sensitivity, and visual acuity of the eye', *Physica*, **10**, 553.

12 Pritchard, R.M., 1961, 'Stabilized images on the retina', *Scientific American*, **204**, 72 (Jun).

13 Goethe, J.W. von, 1970 (reprint), *Theory of Colors*, MIT Press, Cambridge, Mass.

14 Mueller, C.G. and Rudolph, M., 1966, *Light and Vision*, Time-Life Books, New York.

15 Land, E.H., 1959, 'Experiments in color vision', *Scientific American*, **200**, 84 (May).

16 Hurvich, L.M., 1981, *Color Vision*, Sinauer Assoc., Sunderland, Mass.

17 Ratliff, F., 1972, 'Contour and contrast', *Scientific American*, **226**, 90 (Jun).

18 Lynch, D.K., 1991, 'Step brightness changes of distant mountain ridges and their perception', *Applied Optics*, **30**, 3508–3513.

19 Hershenson, M. (Ed.), 1989, *The Moon Illusion*, Lawrence Erlbaum Associates, Hillsdale, N.J.

20 Tolansky, S., 1964, *Optical Illusions*, Pergamon Press, Oxford.

21 Kaufman, L. and Rock, I., 1962, 'The moon illusion', *Scientific American*, **207**, 120.

22 Hochheimer, B.F. and Kues, H.A., 1982, 'Retinal polarization effects', *Applied Optics*, **21**, 3811.

23 Bone, R.A. and Landrum, J.T., 1983, 'Dichroism of lutein: a possible basis for Haidinger's brushes', *Applied Optics*, **22**, 775.

24 White, H.E. and Levatin, P., 1962, 'Floaters in the eye', *Scientific American*, **206**, 119 (Jun).

25 Wood, R.W., 1934, *Physical Optics*, Macmillan, New York.

26 Strong, J., 1958, *Concepts of Classical Optics*, Freeman and Co., San Francisco.

27 Malacara, D., 1978, *Optical Shop Testing*, John Wiley, New York.

28 Bishop, R.L. (Ed.), 1988, *Observer's Handbook*, Royal Astronomical Society of Canada, Toronto.

29 Larmore, L. and Hall, F.F. Jr, 1971, 'Optics for the airborne observer', *Journal of the Society Motion Picture and Television Engineers*, **9**, 87.

30 Wood, E., 1968, *Science for the Airplane Passenger*, Houghton Mifflin, Boston.

8 Exotic clouds

Emerging from a movie house late in the afternoon, eyes blinking from the glare, we happened to glance up toward the nearby Rincon Mountains. There, spanning the summit, was a 'pile-of-plates' cloud display. Perfectly centered, perfectly symmetric, it was an astonishing caprice of Nature.

Observed around 5 PM in March, 1972, at Tucson, Arizona. No photograph.

In Chapter 4, under the section 'Cloudy skies', we described the physical makeup of clouds and how they interact with light. Now we simply enjoy a few cloud photographs as exotic happenings in our atmosphere. Exotic implies 'foreign, unfamiliar, strangely beautiful'. Add to this Minnaert's comment of 'a once in a lifetime experience'. We propose that each of the following photographs matches one of the above meanings of exotic.

Someone, persons unknown to us, coined the term 'pile of plates' as an apt description of the scene in Figure 8.1. This photograph was taken by Chris Gilbert of the British Antarctic Survey at South Georgia Island. Clearly a magnificent sight, but there is more to this cloud than meets the eye. Although appearing absolutely stationary, the cloud's particles are in fact moving at a high speed transverse to the line of sight. Hidden is a mountain top over which this cap is suspended. On the windward side the air is uplifted, cooled, and contained moisture condenses. On the leeward side the water drops evaporate. This presumably accounts for the plate-like layering but we can offer no detailed explanation of the plate-like formation. The lifetime of the cloud drops is measured in seconds, so little or no evolution of the droplets is possible. This means that they are uniform in size, accounting for the iridescence along the thin borders (see cloud banding, Section 4.14).

Similar text could be constructed for the other pictures. Scorer's *Clouds of the World* is full of ideas and conjectures which may be applicable. Honestly, however, we do not really understand these atmospheric events. In these photographs that follow, therefore, the captions are complete without further explanation.

Fig. 8.1 *Orographic altocumulus lenticularis* with irisation, see text. (British Antarctic Survey, photo by C. Gilbert)

Fig. 8.2 Lenticular (lens-like) clouds spanning the Rincon Mountains east of Tucson. The display was striking for the cloud length, singularity, and narrowness. Such clouds always indicate the presence of high velocity winds. Technical: *altocumulus lenticularis*.

Fig. 8.3 In October 1972, this angry turbulent eddy was recorded off the crest of Makalu in Nepal. A Yugoslavian climbing party suffered severe gusts while trying, unsuccessfully, to climb this ridge on the day this picture was taken. Technical: rotor clouds, *orographic altocumulus*. (Photo by D. Vrabec)

Fig. 8.4 Another angry cloud that we have called 'Jaws'. (British Antarctic Survey)

Fig. 8.6 From the perspective of Boulder, Colorado, Foehn conditions (a strong wind from the west) create splendid writhing clouds. Much higher altostratus in the background displays irisation. Technical: *orographic cumulus*.

Fig. 8.5 One careless match produced this giant: a *pyrocumulus* or cumulus induced by a forest fire. In the back country of the San Gabriel Mountains as seen over the 100-inch telescope at Mt. Wilson, California. (Photo by J. Hickox, July 3, 1953)

Fig. 8.7 This phoenix-like formation lasted only a few minutes over the Baboquivari Mountains in southern Arizona. Imagine the wind vagaries to produce this. (Photo by B. Gillespie) Technical: *orographic arched lenticular*.

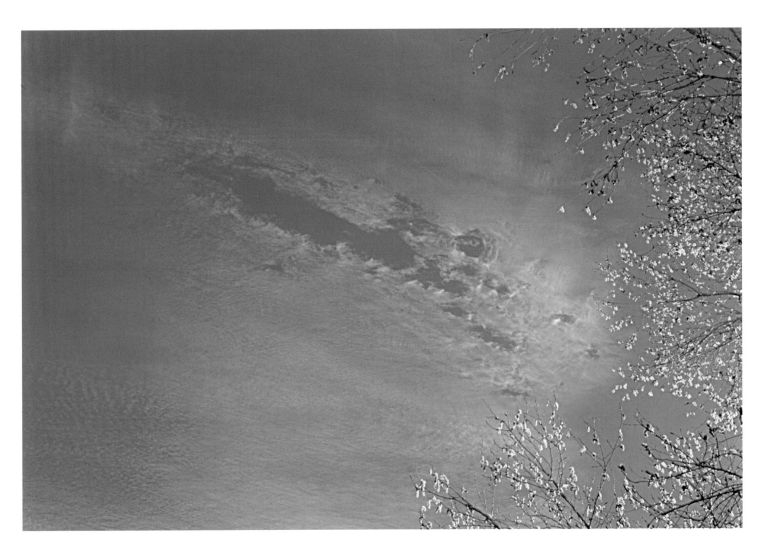

Fig. 8.8 Seen from our own backyard is this curious cirrocumulus layer that has been ruptured by who knows what, a down draft? Technical: *cirrocumulus undulatus lacunosus.*

Fig. 8.9 Tufts of cirrus mimic the desert plants (ocotilla) in the foreground to fulfill the 'strangely beautiful' definition. Technical: *cirrus uncinus*.

Fig. 8.10 A summer thunderstorm spawned these sack-like, or mammary, formations. We are told they indicate strong down drafts in the probable anvil cloud overhead[1]. These drafts project into a more stable layer beneath. Technical: *cumulonimbus mamma*.

Fig. 8.11 Smoke from a notorious Southern California brush fire was the source of this wispy remnant over Pt Dume.

Fig. 8.12 A field of cumulus castellatus over the South China Sea.

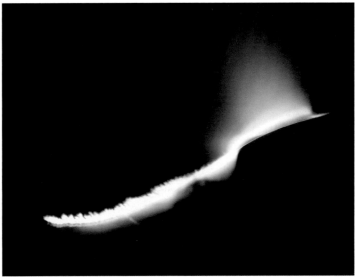

Fig. 8.13 If ever there was an evil cloud this is it: skirt clouds associated with the Soufriere eruption of 17 April, 1979. It is proposed that extremely energetic convection penetrates a moist horizontal layer to uplift it into the vertical[1]. Near instantaneous condensation makes the skirt-like layers visible. Such skirts have been recorded in the lower reaches of nuclear explosion clouds[2].

Fig. 8.15 The southwestern US has recently witnessed a series of rocket-launch clouds. The Hera test rocket is fired near dawn either from the White Sands Missile Range or the old Fort Wingate Army depot close to Gallup, New Mexico. Initial condensation from such a missile produced this spectacular streak with a regular diffraction corona pattern. This was in the morning sky of March 2, 1998 and the observation was from Tucson, a sight-line distance of about 500 km.

Fig. 8.14 In summer the upwelling of cold ocean water along the California coast, together with inland breezes of warm moist air, creates *advection fog*. The pile up of such breaking fog bellows almost duplicate the sea breakers below. At San Simeon, California.

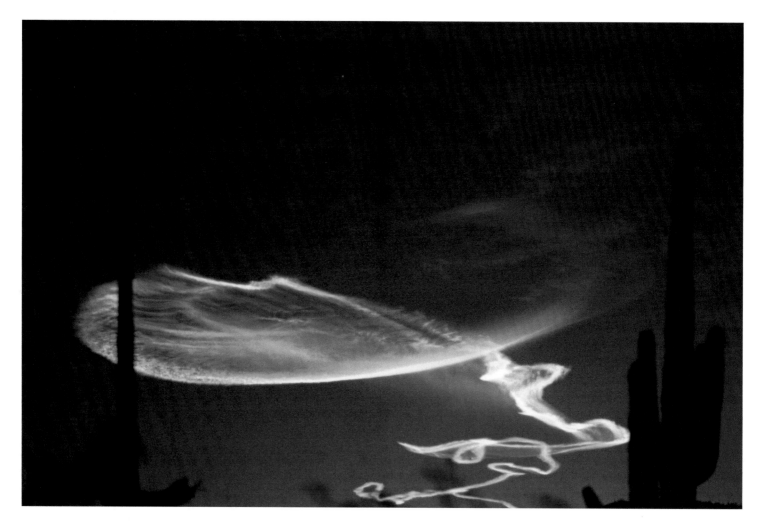

Fig. 8.16 This rocket exhaust trail was blown around by high altitude winds and was visible from Tucson on March 6, 1997. The display of irisation is so vivid because the cloud catches the sun's rays while the local predawn sky is practically dark. (Photo by D. Branston)

References

1 Scorer, R., 1972, p.26 in *Clouds of the World*, Stackpole Books, Harrisburg, Pa.

2 Barr, S., 1982, 'Skirt clouds associated with the Soufriere eruption of 17 April 1979', *Science*, **216**, 1111.

Glossary

A

aberration property of an optical system which leads to an imperfect focus.

absorption the process by which light is retained by matter. See scattering, extinction.

advection predominantly horizontal transport and mixing of parcels of air. Compare convection.

aerosol a suspension of solid or liquid particles in air.

after-glow (1) same as twilight arch. (2) General term for the purple light and the twilight arch. (3) Same as alpenglow.

after-image visual image of brightly illuminated object perceived when eyes are later closed, or when object vanishes.

aguaje same as red tide.

air the mixture of gases that composes the earth's atmosphere.

air mass a measure of the amount of air through which light from outside the atmosphere must pass to reach the observer.

airglow full sky glow at night due to upper atmospheric atomic and molecular emissions.

airlight diffuse, scattered sunlight from air molecules and haze that causes distant objects to lose contrast, becoming bluish or grayish with increasing distance.

albedo ratio of reflected to incident light, especially for a rough or diffusely scattering surface. Snow has an albedo near 1.0 and lamp black near 0.0.

Alexander's dark band relatively dark region of the sky between the primary and secondary rainbows.

almucantar small circle on the celestial sphere centered on the zenith (or nadir) and connecting points of equal altitude.

alpenglow (1) reddish light of mountain peaks when illuminated by the direct low sun. (2) Soft, shadowless, purplish, landscape coloration that results from illumination by the twilight arch after sunset.

altazimuth coordinate system system for specifying the position of an object on the celestial sphere using altitude and azimuth.

altitude angular distance on celestial sphere, measured from the astronomical horizon (altitude = $0°$). The zenith has an altitude of $90°$, the nadir $-90°$.

altocumulus principal cloud type; mostly of water droplets, textured and rounded cloudlets with shading; with coronae/irisation, no halos.

altostratus principal cloud type; water drop layered clouds with shading; no halos.

Andes glow same as Andes lights.

Andes lights St Elmo's fire seen on high mountain peaks.

annular eclipse eclipse of the sun in which the moon fits within the solar disk because its angular size is less than the sun's.

anomalous dispersion dispersion in a wavelength interval where the index of refraction increases with increasing wavelength.

anthelic arcs halo. Any of various pale arcs that cross at the anthelic point.

anthelic pillar halo. Pale, vertical pillar 2–10° high centered roughly on the anthelic point.

anthelic point point on solar almucantar opposite the sun. See solar points.

anthelion halo. Colorless, roughly spherical spot 1–3° across located at the anthelic point.

anticorona same as glory.

anticrepuscular arch same as antitwilight arch.

anticrepuscular rays extensions of crepuscular rays to the hemisphere opposite the sun where they converge to the antisolar point.

antisolar arcs halo. Pale arcs that pass obliquely through the antisolar point.

antisolar point point on celestial sphere 180° from the sun. See solar points.

antitwilight arch pinkish band of scattered sunlight extending to 30° above the earth shadow.

aphakic eye human eye from which the lens has been removed, resulting in enhanced ultraviolet sensitivity.

apparent horizon same as geographic horizon. See horizon.

apparent size angular size of an object. Compare perceived size.

Arago point unpolarized point in the clear sky located on the sun's vertical circle 15–25° above the antisolar point. See neutral points.

arc of contact same as circumscribed halo.

arctic haze low lying, polar haze that reduces horizontal, but not vertical, visibility.

ash gray light same as earthshine.

ashen light same as earthshine.

astronomical horizon plane perpendicular to the zenith and passing through the observer's eye.

astronomical twilight period during which the sun is between altitude −12° and −18°. See twilight (2).

atmosphere the envelope of air surrounding the earth.

aureole (1) bright, colorless glow a few degrees across surrounding the sun (or moon) due to forward scattering by dust, water drops, etc. (2) Any enhancement in the light around the shadow of an observer's head. See heiligenschein, opposition effect.

aureole effect rays observed to radiate from the antisolar point in gently rippled water.

aurora diffuse, transient, slowly moving lights in the high latitude night sky. Usually green or red, often as rays or curtains.

aurora australis an aurora in the southern hemisphere.

aurora borealis an aurora in the northern hemisphere.

auroral arcs long, often curved, aurora with sharp lower edge and diffuse upper portions.

auroral bands long aurora showing kinks or folds.

auroral ribbon an aurora in the shape of a gently folded ribbon.

averted vision technique for viewing a faint object by gazing a few degrees to one side of its position. See scotopic vision, peripheral vision.

azimuth angular measure of the distance from the north point eastward along the horizon to an object's vertical circle. See altazimuth coordinate system.

B

Babinet point unpolarized point in the clear sky located 15–25° above the sun. See neutral points.

backscattering component of scattering towards the hemisphere containing the source of light. Compare forward scattering. See retroreflection.

Baily's beads brilliant points of the solar disk visible around the lunar limb for a few seconds just before and after totality during a total solar eclipse.

ball lightning rare, unexplained balls of light, 2–100 centimeters across, associated with electrical storms.

band of darkness in the clear blue sky, a broad band of minimum brightness and maximum polarization lying on the great circle 90° from the sun. See Rayleigh scattering.

bead lightning lightning stroke that breaks into bead-like chains lasting longer than ordinary lightning strokes.

belt of Venus same as antitwilight arch.

Benard cell any regular, stable pattern of cellular convection.

Bishop's ring pale, usually reddish ring 5–20° across surrounding the sun. Often caused by volcanic dust in the stratosphere.

black aurora dark lanes in a veil aurora.

black ice transparent ice on a landscape feature such as a road.

blackbody theoretically perfect absorber and emitter of electromagnetic radiation.

blind spot visually insensitive location in field of view of the eye where the optic nerve enters the retina.

blind strip in some mirages a featureless, horizontal gap containing only skylight.

blink localized brightening or darkening of a cloud base near the horizon caused by the reflection of sunlight from the surface below. See ice blink, snow blink, water sky, land sky, snow sky, sky map.

blue ice thick, translucent ice displaying blue color.

blue moon rare condition where the moon (sun) appears a pale blue (or green) because of an unusual Mie scattering condition.

blue shadows blue color of shadows, especially when the sun is low.

blue sky blue color of clear air caused by preferential scattering of short wavelength end of visible sunlight by air molecules.

bolide same as fireball.

Bottlinger's rings oval rings centered on the subsun. Not understood.

Bouguer's halo same as fogbow.

Brewster's angle angle of incidence of light on a dielectric surface (e.g. water) that causes the reflected and transmitted rays to be totally polarized.

Brewster point unpolarized point in the clear sky located 15–25° below the sun. See neutral points.

bright glow low sun aureole , extending along the horizon and visible until the solar depression is 7°.

bright segment same as twilight arch.

Brocken bow same as glory when accompanied by the Brocken Spectre.

Brocken Spectre shadow of observer cast onto a mist and, as a result, perceived in depth. See perspective, mountain shadow.

broken corona same as iridescent cloud.

Burney's halo rare, circular, sun-centered halo of radius 19°.

C

capillary wave short wavelength water wave formed by wind and perpetuated by surface tension. Same as ripple.

cardinal point one of four horizontal directions, defined as points on the horizon: north point, east point, south point, west point.

castles in the air same as Fata Morgana.

cat's eyes retroreflection from animal eyes. See retroreflection. Compare heiligenschein.

cat's paws irregular, fleeting dark patches on calm water produced by wind gusts, due to local capillary waves.

caustic (1) three dimensional envelope of imperfectly focussed rays. (2) Two-dimensional pattern formed when a caustic falls on a surface. See light web.

caustic network bright network seen in water on the bottom under a rippled surface, or reflected from that surface onto a wall.

ceiling the lowest level of opaque clouds as seen from below.

celestial equator great circle formed by the projection of the earth's equator onto the celestial sphere.

celestial horizon plane perpendicular to the zenith and passing through the center of the earth. See horizon.

celestial sphere imaginary sphere at infinity surrounding the observer.

Cellini's halo same as heiligenschein.

chain lightning same as bead lightning.

channel lightning same as bead lightning.

chromatic adaptation fatigue of one trichromatic channel to favor another in the eye.

chromosphere reddish outer atmosphere of the sun visible momentarily just before and after totality at a total solar eclipse.

circumhorizontal arc halo. Brilliantly colored horizontal band seen 46–50° below the sun when sun is above 58°.

circumpolar stars stars which do not rise or set during the night but circle the north (or south) celestial poles.

circumscribed halo halo which surrounds the 22° halo and is tangent to it at its upper and lower points.

circumzenithal arc halo. Brilliantly colored crescent 46–50° above the sun and centered on the zenith. Observed when sun is below 32°.

cirrocumulus principal cloud type; a high elevation cloud of small elements without shadows.

cirrostratus principal cloud type; a high elevation fibrous cloud of large extent without shadows.

cirrus principal cloud type; a high elevation, wispy cloud without shadows.

civil twilight period of time when the sun is between 0° and –6° altitude. See twilight (2).

cloud a visible collection of water droplets, ice particles, or dust particles suspended in air.

cloud bow same as fogbow.

cloud contrast bow a circular band of enhanced contrast of cloud structure in the vicinity of the rainbow angle. Compare fogbow.

cloud-to-stratosphere lightning lightning between cumulonimbus and clear air (usually upwards toward the stratosphere).

color interpretation of various wavelengths of light in the visible spectrum by the eye–brain system. See complementary colors, cones, primary colors.

color constancy tendency for objects to be perceived as having their normal or accustomed color independent of illumination hue and brightness.

'color in frost' same as snow sparkles except in frost.

'color in oil-on-water' see interference colors.

'color in snow' (1) same as snow sparkles. (2) Reddish tint of snow due to algae growth.

column ice short, prismatic ice crystal.

complementary colors any two colors which yield white when added together in the proper proportion.

composite flash multiple lightning strokes in the same channel.

condensation the phase transition from vapor to liquid state. Compare evaporation, sublimation.

condensation trail same as contrail.

cones light sensitive cells in the retina used in photopic vision at high levels of light. Compare rods.

conjunction occasion when a planet or the moon has the same right ascension as the sun. Opposite: opposition.

contrail from CONdensation TRAIL. A white cloud streak due to condensation of water vapor initiated by aircraft passage. Compare distrail.

contrast visual degree of brightness or color difference between two adjacent areas.

contrast triangle optical illusion seen as a dark triangle sitting on the horizon above the bright glitter of sunlight (moonlight) over water and below the sun (moon). See Mach phenomena.

convection predominantly vertical transport and mixing of parcels of air. Compare advection.

cornfield effect same as opposition effect.

corona (1) circular bands of color 1–10° in diameter centered on sun (moon) and due to diffraction of light by water drops (or ice). (2) Pale, circular, or oval, glow seen around totally eclipsed sun, part of outer solar atmosphere. (3) Multiple aurorae that appear to converge to a single location in the sky. See aurora, perspective.

corona discharge continuous electrical discharge associated with high voltages in air. See St Elmo's fire.

corposant same as St Elmo's fire.

counter-glow same as gegenschein.

counter-point same as antisolar point.

counter-sun same as anthelion.

counter-twilight same as antitwilight arch.

crepuscle same as twilight.

crepuscular arch same as twilight arch.

crepuscular rays bright or dark rays cast in the atmosphere by clouds, or other objects, that are seen to diverge from the sun. Compare anticrepuscular rays.

cross the simultaneous occurrence of any of several halos that form a cross shape. Usually the pillar and parhelic circle, or the parhelic circle and 22° halo. Same as sun cross.

crown flash hidden flashes from lightning that illuminates the upper parts of cumulonimbus.

culmination highest altitude attained by a celestial object.

cumulonimbus principal cloud type; the thunder shower cloud.

cumulus principal cloud type; an individual rounded, shaded cloud.

cyanometry the study and measurement of the sky's blue color.

D

dark segment same as earth shadow at dawn and dusk.

dawn (1) first detectable light of morning twilight. (2) Twilight period before sunrise.

daylight combined light from sun, sky, clouds and landscape.

declination in the equatorial coordinate system, the angle along an object's hour circle between the equator and the object.

dendrite hexagonal ice crystal with intricate branches.

dew water droplets condensed on grass and other objects which have dropped below the dew point due to radiation cooling. Compare frost, white dew, gutation.

dewbow rainbow in dew-covered grass or on the ground. See horizontal rainbow.

diamond dust small ice crystals (less than 0.1 millimeter) suspended or falling in clear cold air.

Diamond Ring single brilliant flash of sunlight observed just before and after total solar eclipse.

differential refraction the normal vertical distortion of images caused by the different amounts of atmospheric refraction across the object's vertical angular size. See flattened sun (moon).

diffraction the deflection of light around objects arising from its wave properties.

diffuse aurora same as airglow (3).

diffuse reflection reflection by a rough surface which tends to scatter light in every direction. See albedo.

diffusion the process by which energy is transported in matter via molecular collisions. Compare convection.

dip the angle measured downward from the astronomical horizon to the sea-level horizon. Dip is 0° at ground level and increases with the observer's elevation.

dispersion (1) the separation of white light into its spectral colors by redirecting each color into a slightly different direction. (2) The variation of refractive index with wavelength.

dissipation trail same as distrail.

distrail DISsipation TRAIL. A long thin gap in a cloud caused by local evaporation of a cloud by heat from aircraft exhaust.

double rainbow the simultaneous occurrence of both the primary and secondary rainbow.

drapery aurora showing fold-like curtains.

drizzle precipitation whose drop size is less than 0.5 millimeter.

dusk (1) twilight period following sunset. (2) Last perceptible sunlight after sunset. Compare dawn.

dust fine solid particles often present in air.

Dutheil's halo rare, circular sun-centered halo of radius 24°.

E

earth shadow dark, bluish region of the sky opposite the setting sun where the earth's shadow is cast upon the atmosphere.

earthlight same as earthshine.

earthshine dark side of moon illuminated by sunlight reflected from earth.

east point cardinal point on the horizon directly east, azimuth = 90°.

eclipse of the moon same as lunar eclipse.

eclipse of the sun same as solar eclipse.

ecliptic great circle defined by the annual path of the sun against the background stars.

electromagnetic spectrum the entire range of electromagnetic waves, including radio waves, infrared radiation, visible light, ultraviolet light, x-rays, and gamma rays.

electrometeors any electrically-induced light in the sky such as lightning, ball lightning, St Elmo's fire.

elevation height above sea level.

elongation angular distance between the sun and a solar system body as observed from the earth.

elves pancake shaped flashes high above thunderstorms. Compare sprites.

emission any process that converts energy into light. Compare absorption.

entopic halo apparent diffuse illumination around any isolated bright light observed at night due to scattering within the eye.

epoch (1) a fixed point in time. Instant of an observation. (2) Time or date when coordinates are valid.

equatorial coordinate system celestial coordinate system used by astronomers that is fixed with respect to the background stars and aligned with the earth's rotational axis.

equinox (1) either of two points on the celestial sphere representing the sun's position as it crosses the celestial equator. See vernal equinox, autumnal equinox. (2) The moment in time when the sun crosses the celestial equator.

evaporation change from liquid to vapor state. Compare condensation, sublimation.

evening star brilliant star-like object in the western twilight, usually Venus. See morning star.

extended source (object) source of light that has a perceptible angular size (e.g. sun or moon). Extended sources form shadows with both umbrae and penumbrae. Compare point source.

extinction the attenuation of a light beam due to absorption and/or scattering. See absorption, scattering.

F

falling star same as meteor.

fallstreak hole hole in thin altocumulus due to virga-like precipitation.

fallstreaks wispy, near vertical clouds attached to and extending below altocumulus or cirrocumulus. Compare virga, precipitation trails.

false dawn zodiacal light in the eastern sky before morning twilight.

Fata Bromosa nearly featureless superior mirage resembling a fog bank. See mirage. Compare Fata Morgana.

Fata Morgana superior mirage characterized by irregular sharp features seen over water and cold ground. See mirage, Hafgerdingar effect. Compare Fata Bromosa.

ferruginous water water with a metallic-like scum producing an interference reflection.

Feuille's halo rare, circular sun-centered halo of 32° radius.

fireball unusually bright meteor, especially one that explodes.

firnspiegel thin layer of ice over snow which reflects light. Literally 'ice mirror' in German.

flaming aurora aurora display with flickering and flashing of brightness on time scale of a second.

'flattened sun (moon)' vertical distortion of the low sun caused by normal atmospheric differential refraction.

floaters small, slowly moving, indistinct spots in the visual field of view due to out-of-focus retinal cells in the eye's humor.

fluorescence conversion of shorter wavelengths of light into longer wavelengths via absorption and re-emission.

fog cloud at, or near, the ground.

fogbow pale, often colorless rainbow in fog.

fog-eater name for fogbow by Nova Scotia fishermen.

forked lightning lightning flash in which one or more branches depart from the main discharge channel.

forward scattering light scattering which occurs into the forward hemisphere. Compare backscattering.

fovea central region of the retina where photopic vision is most acute. See photopic vision.

Fraunhofer line dark, narrow wavelength feature in the spectrum of sunlight due to light absorption by atoms or molecules in the terrestrial or solar atmosphere.

frost ice sublimed on surfaces whose temperature is below both the air's dew point and freezing point.

G

Galle's arc same as halo of 46°.

gegenschein faint, nearly circular patch of light in the night sky located at the antisolar point.

geographic horizon distant boundary between sky and earth. Our usual concept of horizon.

geometric horizon plane separating sky from sea for a sea-level observer in the absence of refraction. See horizon.

geostationary satellite geosynchronous satellite whose orbital is such that it remains in a fixed position in geographic coordinates.

geosynchronous satellite artificial satellite that orbits the earth in 24 hours.

'Gibson girl' imagined face in the full moon. See man in the moon.

glaciation conversion of water drops to ice crystals in clouds.

glare ice same as black ice.

glaze same as black ice.

glitter sparkling patch of reflected sunlight on rippled water between observer and the sun. See moon circles.

glitter path same as glitter.

globe lightning same as ball lightning.

gloom exceptionally dark, gray, overcast landscape illumination.

glory colored circular bands 2–10° across around the antisolar point on clouds.

golden bridge same as glitter.

gravity wave wave whose restoring force is gravity, as in ocean waves. Compare capillary wave.

great circle any circle in the sky that divides the celestial sphere into two equal hemispheres.

green flash the last segment of the setting sun which momentarily appears green. See green rim.

green ray same as green flash.

green rim upper segment of the low sun.

green shadow rare greenish coloration of shadows.

ground flash cloud to earth lightning flash.

gutation water drop exuded by foliage under humid conditions.

H

Hafgerdingar effect a superior mirage with irregular looming, literally 'sea fences' in Icelandic.

Haidinger's brushes a low contrast cross about 5° in width with orthogonal yellow and blue lobes seen when viewing a low contrast scene in polarized light. Physiological in origin.

Hall's halo same as Rankin's halo.

halo any of various rings, arcs, and spots in the sky due to reflection and refraction of sunlight (or moonlight) in atmospheric ice crystals.

halo of 22° common, sun-centered halo of 22° radius.

halo of 46° relatively uncommon sun-centered halo of 46° radius.

'hare in moon' imagined animal figure on face of the full moon, especially common in oriental lore.

harvest moon the occurrence of the nearly full, slow-rising moon around the time of autumnal equinox. See moon illusion, hunter's moon.

haze an aerosol whose particles are small enough to introduce a slight coloration.

heat lightning lightning too distant for accompanying thunder to be heard.

heiligenschein colorless glow in the landscape around the antisolar point due to backscatter from resting water drops. See opposition effect.

Hesperus same as evening star.

Hevel's halo rare, circular, sun-centered halo of 90° radius.

Hevel's parhelia halo. Rare parhelia located at 90° solar azimuth.

Hevelius' halo same as Hevel's halo.

highway mirage common inferior mirage seen over a highway.

Hillingar effect weak superior mirage over large area that makes the earth appear flat or concave upwards.

'holes in cloud' same as lacunosus.

horizon any of various divisions between the sky and the earth or sea. See local horizon, geographic horizon, sea-level horizon, astronomical horizon, celestial horizon.

horizontal rainbow same as dewbow.

hour angle angle at the celestial pole between the hour circle of an object and the meridian.

hour circle great circle passing through an object and the celestial poles.

hunter's moon full moon one month after the harvest moon.

hydrometeor any form of water or ice falling from the sky.

hyperopic vision farsighted. Compare myopic vision.

I

ice solid state of water. See column ice, ice crystal, plate crystal, rime, snowflake, frost, snow, black ice, glare ice, diamond dust, glaze, whiteout.

ice blink bright undersides of clouds due to light reflected from ice-covered ground. See blink.

ice crystal single crystal of frozen water.

Ignis Fatuus same as will o' the wisps.

index of refraction measure of a substance's ability to refract light. See dispersion.

induced color tendency for the perceived color of an object to depend on surrounding hues.

inferior mirage mirage in which the refracted image is below the object's true position. Compare superior mirage.

infralateral arcs halos. Any of various arcs associated with the 22° and 46° halos, lying generally on either side of the sun.

infrared region of the electromagnetic spectrum just beyond the red end of the visible spectrum.

insolation INcoming SOLar radiATION. The variable amount of electromagnetic radiation received from the sun at the earth's surface. Compare solar constant.

interference constructive or destructive interaction of light waves.

intracloud lightning lightning discharge inside or between clouds.

inversion see temperature inversion.

ion electrically charged atom or molecule.

ionosphere atmosphere above 70 kilometers characterized by a high percentage of ions.

iridescence color due to the interference of light. See corona (1), glory.

iridescent cloud cloud displaying an irregular corona, or fragments thereof.

irisation same as iridescence.

irradiation the apparent enlargement of a bright object because of scattering in the air, eye, or photographic film. See entopic halo.

J

Jacob's ladder same as crepuscular ray.

K

Kage Fuji triangular shadow of Mt Fuji as seen from the summit. See mountain shadow.

Kern's arc rare halo, probably a reflected circumzenithal arc.

L

lacunosus layer cloud characterized by containing nearly circular holes.

land sky relative darkening on the underside of low clouds that indicates the presence of otherwise invisible land. See water sky, sky map, blink.

landpools horizontally elongated 'images' of a landscape on gently rippled water. Compare skypools.

lateral arcs any of various halos found on either side of the halo of 22° or halo of 46°.

lateral mirage inferior mirage on a vertical surface where image of an object is displaced horizontally. See mirage.

laurence same as scintillation.

Lenard effect separation of electrical charge associated with the aerodynamic break-up of falling water drops.

lenticular cloud lens-shaped orographic clouds. See lee waves, orographic clouds.

light visible electromagnetic radiation.

light chaos random sensations of light in complete darkness.

'light of the night sky' all light that prevents the night sky from being totally dark e.g. starlight and airglow. See urban glow, aurora.

lightning bright flashes of light due to electrical discharges in thunder storms. See ball lightning, bead lightning, sheet lightning, heat lightning, ribbon lightning, Lenard effect.

light pillar same as pillar.

limb visual edge of an astronomical body.

limb darkening the decrease of intensity of the solar disk towards its limb.

line lightning lightning free of dendritic arms.

line-of-sight path by which light from an object reaches the eye.

lithometeor visible collection of dry solid material in the sky such as dust or smoke.

local horizon boundary between the sky and upper outline of terrestrial objects. See horizon.

Local Standard Time mean solar time within a geographic zone of longitude; clock time.

looming mirage in which an object appears above its true position. Compare sinking.

LOS Line-Of-Sight.

Lowitz arcs lateral arcs of the 22° halo.

luminescence any non-thermal emission of light.

luminous night cloud same as noctilucent cloud.

lunar eclipse phenomena that occurs when the full moon orbits into the earth's shadow and darkens.

lunar rainbow rainbow formed by moonlight.

lunar twilight increase of sky brightness due to the presence of the nearly full moon below the horizon.

M

Mach bands optical illusion appearing as light and dark bands when an otherwise featureless field contains a non-uniform intensity gradient to dark.

magnitude measure of an astronomical source's brightness on a logarithmic scale.

man in the moon commonly imagined face in the full moon, especially common in occidental lore. See Gibson girl, hare in the moon.

marine rainbow rainbow seen in ocean spray.

mean solar time time reckoned by a fictitious sun which moves uniformly eastward along the celestial equator (1 mean solar day = $24^h 0^m 0^s$). See sun dial time, sidereal time, Universal Time, Local Standard Time.

meridian vertical circle passing through the zenith, nadir, north and south points.

mesolateral arc rare halo seen as a short arc passing vertically through the parhelion position

mesosphere atmospheric layer between the stratosphere and thermosphere (20–80 kilometers, mid-latitudes).

mesospheric clouds same as noctilucent clouds.

meteor fast moving point of light in the clear night sky arising from the frictional heating of a meteoroid entering earth's upper atmosphere. See meteoroid, meteorite, bolide, meteor shower.

meteor shower usually predictable swarms of meteors originating from a common point in the sky (radiant).

meteor trail visible trail of self-luminous, ionized gas along the trajectory of a meteor.

meteorite meteor which reaches the ground before disintegrating.

meteoroid interplanetary rock or dust particle before entering the atmosphere to become a meteor.

meteorological optics the study of optical effects that occur naturally in the landscape and sky.

meteorological range the distance at which the contrast of an extended black object near the horizon is reduced to 2%. See visual range.

micrometer one millionth (10^{-6}) of a meter.

midnight sun the presence of the sun at local midnight in polar regions (latitudes exceeding 67.5°) during summer.

Mie scattering light scattering by spherical particles. Compare Rayleigh scattering. See scattering.

milky sea even, diffuse, nocturnal bioluminescent glow in surface water extending to great distance.

milky weather same as whiteout.

minimum deviation angle the smallest angle through which light is deviated in passing through a dispersing element.

mirage any atmospheric image not normally present that is caused by refraction in an abnormal temperature gradient. See superior mirage, inferior mirage, lateral mirage, Novaya Zemlya, Fata Morgana, Fata Bromosa, towering, looming, stooping, sinking.

mirror image virtual image formed by a smooth, flat, reflecting surface.

mist fine water drops suspended in the air and less dense than a fog. Compare fog.

mist bow same as fogbow.

mock mirage inverted narrow stripes of the solar image above the sun at sunset.

mock sun same as parhelion.

moon bow same as lunar rainbow.

moon circles closed loops of reflected moonlight on gently rippled water. See glitter.

moon dog same as paraselene.

moon illusion optical illusion in which the full moon (or sun) seems larger near the horizon than when overhead. See optical illusion.

moon pillar same as pillar.

moonlight sunlight reflected from the moon that illuminates the earth and landscape.

morning star brilliant point of light in the morning twilight, usually Venus. See evening star.

mother-of-pearl clouds colorful, iridescent clouds seen against the twilight sky, usually nacreous clouds.

'mountain shadows' triangular appearance of mountain shadows when viewed from their summit. See perspective.

mountain shadow spike dark upward-extending spike at the apex of a mountain shadow when viewed from off the summit. See mountain shadow.

'moving ripples in clouds' rarely observed pattern of rapidly moving corrugations in cloud layers.

multiple scattering scattering process in which light undergoes more than one scattering. See scattering. Compare single scattering.

multiple suns (moons) unexplained occurrences of multiple images of the sun.

myopic vision nearsightedness. Compare hyperopic vision.

N

nacreous clouds twilight clouds seen during winter months at high latitudes.

nadir point on the celestial sphere opposite the zenith.

nanometer one billionth (10^{-9}) of a meter.

nautical twilight period of time when the sun is between 0° and −12° altitude. Compare astronomical twilight, civil twilight.

neutral points any of several locations in the clear sky on the solar vertical circle where the polarization is zero. See Arago point, Babinet point, Brewster point.

nightglow same as airglow.

noctilucent clouds high, bluish-white clouds seen in the twilight particularly in summer.

nocturnal twilight twilight that lasts all night in summer at high latitudes because the sun is just below the horizon.

non-polar aurorae same as airglow.

north celestial pole point on the celestial sphere about which the heavens appear to rotate once each day; defined by the projection of the geographic north pole onto the meridian.

north dawn same as aurora borealis.

north point cardinal point on the horizon due north at azimuth = 0°. See cardinal points.

northern lights same as aurora borealis.

Novaya Zemlya a superior mirage involving long range optical ducting over polar terrain.

O

oblique anthelic arcs same as anthelic arcs.

oblique heliacal arcs halo. Colorless arcs that pass obliquely through the anthelic point.

occultation the passage of one astronomical body across another and thereby blocking it from view.

okra a tall annual (*Hibiscus esculentus*) widely cultivated in the southern US and West Indies for its mucilaginous green pods that are pickled or used as the basis of soups or stews. Not associated with any known optical phenomenon.

okta (British) octal unit describing sky cover. Complete overcast is 8 oktas.

Olbers' paradox the question of why the night sky is not as bright as the sun if the universe is infinite and filled with an infinite number of stars.

opposition time when a planet or the moon has a right ascension 12 hours different from the sun. Opposite: conjunction.

opposition effect diffuse bright patch 1–2° across at the antisolar point in the landscape. See shadow hiding, heiligenschein, retroreflection.

optical depth measure of opacity on logarithmic scale. Transparent objects have zero optical depth, translucent systems (thin haze) have optical depths between zero and 4 and opaque objects have optical depths exceeding 4 (less than 2% transmission).

optical illusion any of various optical effects that cause the conscious mind to interpret an image incorrectly. See moon illusion, Mach bands, contrast triangle.

optical manhole the 180° celestial hemisphere compressed by refraction into a cone of 96° across to an observer under water.

P

parafoveal vision photopic vision away from the fovea. Compare averted vision.

parallax the apparent displacement, or the difference in position, of a nearby object against the background as seen from two viewing locations.

paranthelic arcs halo. Any arc which passes through the positions of the paranthelia.

paranthelion halo. Either of two whitish patches of light a degree or so across found at the solar elevation and at an azimuth of 120° relative to the sun.

paraselene halo. Parhelion formed by moonlight.

parhelic circle halo. Colorless, horizontal band which encircles the sky at the solar altitude.

parhelion halo. Either of two bright, usually colorful patches of light found at the solar altitude on, or just outside the 22° halo.

Parry arcs halo. Either of two arcs found above and below the position of the 22° halo.

partial eclipse eclipse when only part of the sun or moon is hidden, or any part of an eclipse except total.

pearl lightning same as bead lightning.

penumbra region of a shadow where the light source is partially covered. See shadow. Compare umbra.

perceived size impression of physical size in, say, meters. Compare apparent size.

peripheral vision vision near the edge of the eye's field of view. Compare averted vision.

permanent aurora same as airglow.

perspective apparent geometrical distortion of objects in three dimensions when projected onto a two-dimensional surface (e.g. landscape onto a film plane of a camera or the retina of the eye). See mountain shadows, searchlight effect, crepuscular rays, radiant.

phosphenes sensations of light in the eye produced in darkness by an external pressure. Compare light chaos.

phosphorescence light caused by the absorption of radiation and which continues for a time after the exciting source has stopped.

phosphorescent water any of various glows in disturbed water at night, usually due to light-emitting microorganisms (actually chemiluminescence). See red tide.

Phosphorus same as morning star.

photon basic, discrete element of electromagnetic energy.

photopic vision high light level vision employing retinal cones and providing the sensation of color.

photosphere visible surface of the sun in white light.

pillar halo. A vertical shaft of light extending above or below the sun, moon, or other light source.

pilot's bow complete circular rainbow observed in rain from the vantage point of an airplane.

pinhole effect the ability of tiny apertures to produce images without lenses. See squinting.

plate crystal flat, hexagonal ice crystal responsible for many halos.

point location on the celestial sphere, i.e. a direction in space.

point source source of light without any perceptible angular size, for example a star. Compare extended source.

polarization (1) plane of the electric field component of electromagnetic radiation. (2) Process of bringing about a preferred state for (1).

pollen coronae coronae produced by diffraction by pollen particles.

praecipitatio same as precipitation trail.

precipitation any water or ice particles that fall from the sky.

precipitation trail rain sheets extending from a cloud to the ground. Compare fallstreaks, virga.

primary color any one of three broad band colors whose mixture with the other two can produce white and all colors of the spectrum.

primary rainbow the common rainbow of angular radius 42°. See secondary rainbow.

projection mirage same as superior mirage.

prominence reddish 'cloud of gas' sometimes seen protruding just outside sun during a total solar eclipse.

pulsating arcs aurorae which pulsate in brightness with periods of a few seconds to a few minutes.

Purkinje effect blueward shift of the color sensitivity of the eye in dim light.

purple light an infrequent pinkish light in the twilight sky above the solar point when the latter is between –2° and –6°.

Putnin's arcs halo. Same as Parry arcs.

Q

quadrature celestial configuration of a moon or planet in which the sun, earth (or planet) and the object form a right angle. Compare opposition.

quantum same as photon.

R

radiant the location on the celestial sphere where meteor showers seem to originate. See perspective.

radiation (1) electromagnetic radiation. (2) Energy transport by electromagnetic radiation. Compare convection, diffusion.

rain precipitation of water drops whose diameters exceed 0.5 millimeters.

rain streaks same as precipitation trails.

rainband region of the visible spectrum in the yellow where water vapor absorbs light. See Fraunhofer line.

rainbow colored, circular band of light, 2° across, centered on the antisolar point and lying 42° away (primary rainbow) or 51° away (secondary rainbow).

Rankine's halo rare, circular sun-centered halo of radius 17°.

Rayleigh scattering scattering by air molecules, or other particles small compared to the wavelength of light. Compare Mie scattering.

rays shafts of light seen by scattered light from hazy air. See crepuscular rays, anticrepuscular rays, aurora borealis (australis), aureole effect.

'rays in water' same as aureole effect.

Rays of Buddha same as crepuscular rays.

red arcs reddish auroral arcs.

red rim reddish lower limb of low sun. Compare green rim.

red snow pinkish tint in snow caused by algae.

red tide red-brown color of ocean water produced by micro organisms. Compare phosphorescent water.

reflection optical interaction of light with matter in which the incoming ray's directional component perpendicular to the surface is reversed.

reflection rainbow rainbow arising from sunlight reflected from a body of water.

refraction the change in direction of light as it passes obliquely into a medium with a different index of refraction (e.g. from air to water).

refractive index same as index of refraction.

retroreflection any of various optical mechanisms that reverse the direction of light. See heiligenschein, opposition effect.

ribbon lightning rapid successive lightning strokes that appear to shift sideways between strokes because of wind.

right ascension in equatorial coordinates, the angle between an object's hour circle and the vernal equinox. Orthogonal to declination. Compare hour angle.

rime ice deposited on an object when struck by supercooled water drops.

'road to happiness' same as glitter.

rods retinal cells necessary for low light level vision. See scotopic vision. Compare cones.

'Ropes of Maui' same as crepuscular rays.

rosy zone same as purple light.

S

St Elmo's fire steady electrical discharge from objects during electrical storms.

scattering the interaction and redirection of light by any particle or surface. See Rayleigh scattering, Mie scattering, single scattering, multiple scattering.

Scheiner's halo rare, circular sun-centered halo of radius 28°.

schlieren (1) parcels of air whose density variations are revealed by refraction through them. (2) The patterns produced by schlieren.

scintillation rapid brightness, color, and slight position changes of stars or other distant point sources due to refraction by constantly moving air parcels. See seeing, schlieren.

scotopic vision low light level vision employing only retinal rods. Compare photopic vision.

sea glint glitter from sea or ocean surface, especially as seen from an aircraft or space.

sea-level horizon apparent boundary between sky and sea. See horizon.

'searchlight effect' apparent abrupt termination of a searchlight beam when the line-of-sight parallels the beam direction. An illusion of perspective.

secondary rainbow rainbow of 51° radius centered on the antisolar point.

seeing a visual measure of the optical steadiness of the air, usually judged by looking at stars. Compare scintillation, twinkling, schlieren.

self-centered rays same as aureole effect.

shadow (1) volume of space completely, or partially, shaded by a body in front of a light source. (2) The visible presence of the shadow, usually on a solid surface, but occasionally on an aerosol. See mountain shadow, earth shadow, umbra, penumbra, spikes, Brocken Spectre.

shadow bands low contrast, moving bands of light on the ground a few minutes before, and after, a total eclipse of the sun. Also seen on sunlit surfaces when only a small part of the low sun's disk is visible because of obscuration. See schlieren.

shadow hiding the absence of shadows at the antisolar point which produces a bright spot. See opposition effect.

shadow of the earth same as earth shadow.

sheet lightning clouds illuminated by lightning that is obscured. See crown flash. Compare heat lightning, summer lightning.

shimmer seeing effects on extended objects. See seeing, schlieren, scintillation.

shooting star same as meteor.

sidereal same as stellar.

sidereal period time of revolution of a body with respect to the stars. Compare synodic period.

sidereal time astronomical measure of time based on the locations of stars and defined as the hour angle of the vernal equinox (1 sidereal day = $23^h 56^m 4^s$). Compare mean solar time.

silver lining brilliant, white rim of a cloud near the sun, caused by forward scattering.

single scattering scattering process in which light undergoes scattering one time. See scattering. Compare multiple scattering.

sinking mirage in which an object appears below its true position. Opposite of looming. Compare stooping.

sky cover (1) type of cloud obscuring the clear sky. (2) The percentage of cloud coverage.

sky map variable brightness on the underside of cloud decks caused by differing reflectivities of the earth's surface below. See water sky, land sky, blink.

skylight any indirect sunlight from the sky or clouds.

skypools horizontally elongated 'images' of reflected skylight on the surface of gently undulating water. Compare landpools.

sleet mixture of rain and snow.

slicks relatively calm patches of water surrounded by small surface tension ripples and caused by a thin layer of oil. Compare cat's paws.

small circle any circle on the celestial sphere that does not divide the celestial sphere into two equal halves. Compare great circle.

smog SMoke and fOG. Bluish, or brownish haze caused by photochemical aerosols released by automobile exhaust and other industrial pollutants.

smoke combustion-generated aerosol.

snow (1) precipitation consisting of ice particles less than 5 millimeters across, or aggregates of such particles. (2) The accumulation of ice particles of all types on the ground.

snow blink same as ice blink except brighter.

snow sparkles (1) colorless points of light produced by specular reflections from ice crystals on the ground. (2) Colorful points of light seen on snow due to dispersion of light in single ice crystals. See color in snow, color in frost.

snowflake single or aggregate snow crystal.

solar aureole same as aureole (1), (2).

solar constant the amount of solar radiative energy incident at the top of the atmosphere. Compare insolation.

solar corona same as corona (2).

solar eclipse eclipse of the sun by the moon. Compare lunar eclipse. See Baily's beads, Diamond Ring, shadow bands (1), chromosphere, corona (2).

solar point location of the sun's center on the celestial sphere.

solar points any of four locations on the celestial sphere along the solar vertical circle determined by the location of the sun relative to the horizon. The four points are: (1) solar point located at the sun's center; (2) the subsolar point located at the solar azimuth and at the negative solar elevation; (3) the antisolar point located opposite the sun; and (4) the anthelic point located at the solar elevation and at 180° azimuth relative to the sun.

solar radiation electromagnetic radiation received from the sun outside the earth's atmosphere. Compare sunlight, daylight.

solstice (1) either of two points on the ecliptic where the sun is at its maximum and minimum declinations. (2) The instant or season when the sun reaches its maximum or minimum declinations. See winter solstice, summer solstice.

south celestial pole projection of the south pole onto the celestial sphere. See north celestial pole, celestial sphere.

south dawn same as aurora australis.

south point cardinal point on the horizon directly south, azimuth = 180°. See celestial sphere, cardinal points.

southern lights same as aurora australis.

spectre bow same as Brocken bow.

specular reflection reflection from an optically smooth surface such as a mirror or water drop. Compare diffuse reflection.

spike same as mountain shadow spike.

sprites high altitude luminous flashes above thunderstorms. Compare elves.

spurious bows same as supernumerary rainbows.

squinting the act of partially closing the eyes so as to sharpen an image. See pinhole effect.

stephanome an instrument for measuring the angular size of fogbows and halos.

stooping mirage in which objects appear demagnified in the vertical direction. Opposite of towering.

stratosphere atmospheric layer above the troposphere and below the mesosphere.

stratospheric clouds unusual clouds in the stratosphere, for example nacreous clouds, ultracirrus.

stratus principal cloud type; a low elevation, uniform layer cloud.

streaks on water slicks.

subanthelic point same as antisolar point.

sublimation change of state directly from vapor to solid, or its reverse. Compare condensation, evaporation.

subsolar point the point on the celestial sphere at the solar azimuth and negative solar altitude.

subsun halo. Bright colorless spot found at subsolar point on the celestial sphere. See Bottlinger's rings.

summer lightning same as heat lightning.

summer solstice the most northerly declination reached by the sun, occurring approximately on June 21.

sun big, bright, ball in the sky, visible during the daytime in clear air by people out of doors who are not blind.

sun beam same as crepuscular ray.

sun cross same as cross.

sun dog same as parhelion.

sun drawing water same as crepuscular ray.

sun pillar same as pillar.

sundial time approximate time reckoned by a sundial.

sunlight radiation from the sun at ground level, i.e. as modified by the earth's atmosphere.

sunspot dark feature on the solar disk.

sunstreak same as sun pillar.

sunstreaks same as crepuscular ray.

supercooled water liquid water whose temperature is below its freezing point.

superior mirage mirage in which the refracted image appears above its true position. See Hafgerdingar effect, Hillingar effect, Fata Bromosa, Fata Morgana, Novaya Zemlya.

supernumerary rainbows bands of faintly colored light lying just inside the primary rainbow.

supersaturated air air whose relative humidity exceeds 100%.

supralateral arcs halos. Any of several arcs associated with the 22° halo and 46° halo and lying outside these halos.

surf mirage towering or looming of waves on the horizon as seen with binoculars.

surface tension molecular force at the surface of a liquid tending to minimize the exposed area.

surface tension wave same as capillary wave.

synodic period time of revolution on a body with respect to the earth. Compare siderial period.

syzygy astronomical configuration in which the earth, moon and sun all lie in nearly in straight line.

T

tangent arcs halos. Any of various arcs tangent to the 22° halo or 46° halo.

temperature inversion atmospheric layer where the sign of the temperature gradient is reversed to the normal gradient.

terminator boundary line of a planet's disk between daylight and darkness.

tertiary rainbow rainbow arising from three internal reflections (seen in the laboratory but not in nature). Compare primary rainbow, secondary rainbow.

thermosphere atmospheric layer extending from the mesosphere to outer space (80 kilometers outwards).

total internal reflection reflection of light inside a medium at an interface in which all of the light is reflected and none transmitted or absorbed.

towering mirage in which objects appear magnified in the vertical direction. Opposite from stooping.

triboelectric lightning lightning in a volcanic ash cloud caused by grain collisions.

troposphere lowest region of the atmosphere in which most weather and clouds occur.

true horizon same as celestial horizon. See horizon.

turbulence air flow characterized by random velocity and density fluctuations.

twilight (1) region of sky illuminated by the sun which is below the horizon. (2) Period of time when the sun is below the horizon but still illuminates the sky. See astronomical twilight, nautical twilight, civil twilight.

twilight airglow airglow observed during twilight.

twilight arch the glow along the western horizon following sunset when the solar point is 7–18° below the horizon.

twilight wedge same as earth shadow.

twinkling same as scintillation.

U

UFO unidentified flying object.

Ulloa's bow same as glory.

Ulloa's ring same as fogbow.

ultracirrus post sunset, thin, high elevation clouds seen at times of volcanic activity.

ultraviolet light electromagnetic radiation with wavelengths shorter than 400 nanometer.

umbra region of shadow into which no direct light falls. See shadow.

Umov effect tendency of dark objects to show greater polarization than light objects.

undersun same as subsun.

Universal Time mean solar time at Greenwich (0° longitude). See mean solar time. Compare sundial time, sidereal time, Local Standard Time.

urban glow city lights indirectly visible by reflection from clouds or scattering from the atmosphere.

V

van Buisen's halo rare, circular, sun-centered halo of radius 8°.

vanishing point any point on the celestial sphere to which parallel lines appear to converge. See perspective.

vapor trails same as contrails.

veil aurora nearly uniform aurora covering much of the sky.

vertical circle any great circle passing through the zenith.

vertical rainbow (apparently) linear fragment of a reflection rainbow.

virga vertical streaks below a precipitating cloud which do not reach to the ground. Compare fallstreaks, precipitation trails.

visibility same as visual range.

visible light range of the electromagnetic spectrum easily sensed by the normal human eye, generally between 400 and 700 nanometer.

visual range a measure of the clarity of air equal to the largest distance at which extended objects can be seen. See meteorological range, extended source.

vog VOlcanic smoG. Natural aerosol pollutant produced by volcanic emissions.

volcanic dust fine particles injected into the atmosphere by a volcanic eruption.

W

water sky the underside of clouds which appear dark and thus reveal the presence of unseen water in the distance. See blink.

watermelon snow same as red snow.

Wegener arcs halo. Any of various arcs passing through the anthelic point.

west point cardinal point on the horizon directly west, azimuth = 270°. See celestial sphere.

'wet spot' darkening of a wet surface compared to its dry appearance.

white distribution of visible light that produces a white color response in the eye.

white dew dew that formed as water and then froze. Compare frost, dew.

white light light composed of all the colors of the visible spectrum which together appear white to the eye.

white night same as nocturnal twilight.

white rainbow same as fogbow.

whiteout a contrast condition in a snowy, cloudy landscape where the entire visual field is a featureless white. See contrast.

will o' the wisps pale, shimmering, flame-like lights seen over swamps and bogs.

winter solstice the most southerly declination reached by the sun on approximately December 21. Compare summer solstice.

Z

zenith point on celestial sphere directly overhead. Compare nadir.

zenith distance angular distance between the zenith and a point on the celestial sphere.

zodiacal light faint broad band of light extending along the ecliptic from the horizon and visible in the western sky for a few hours following sunset or eastern sky preceding sunrise. See false dawn, gegenschein.

zombie mental state of a person after compiling a glossary.

Index

Specifically named sections are in **bold**.

aberration 259
absorption 21, 259
advection 259
aerosol 259
after-glow 32, **41**, 259
after-images 234, 259
aguaje 259
air 21, 259
airglow **63**, 224, 259
airlight **29**, 259
air mass 23, 259
airplane, observations from 109, 244, 256
albedo 216, 259
Alexander's dark band 112, **122**, 259
Algol 223
almucantar 259
alpenglow 32, **41**, 259
altocumulus 259
altostratus 259
Andes glow 151, 259
angle measurement **239**
annular eclipse **205**, 260
anomalous dispersion 260
anthelic arcs 260
anthelic pillar 260
anthelion **185**, 260
anticrepuscular rays 16, 18, 260
antisolar arcs 260
antisolar point 34, 260
antitwilight arch **38**, 260
aphakic eye 230, 232, 260
apparent horizon 24, 47, 260
apparent size 239, 260
Arago point 27, 260
ashen light 214, 260
astronomical twilight 34, 260
atmosphere 21, 260
atmospheric absorption **21**

atmospheric refraction **46**
  abnormal 46, 54
  normal 46
  wavelength dependence 46
aureole 24, **32**, 260
aureole effect 102, 260
aurora borealis (australis) **63**, 244, 260
auroral arcs 63, 260
auroral bands 64, 260
auroral oval 63, 69
auroral ribbon 63, 260
averted vision 232, 260

Babinet point 27, 260
backscatter in landscape 33, 260
Baily's beads 200, 260
ball lightning 151, 260
band of darkness 26, 260
bead lightning 151, 260
belt of Venus 260
Benard cell 260
Bishop's ring 33, 260
black aurora 260
blackbody 260
black ice 260
blind spot 230, 260
blind strip 260
blinks **144**, 260
  ice 144, 264
  land 144, 265
  sky map 144, 269
  snow 146, 269
  water sky 145, 271
blue moon (sun) **149**, 260
blue sky **22**, 26, 29, 261
bolide 217, 261
Bottlinger's rings **183**, 261
Bouguer's halo 125, 261
Bravais' arc 176

Brewster's angle 72, 261
Brewster point 27, 261
bright glow 261
bright segment 36, 261
brightness 21
  color 21, 39
  polarization 26
  zenith 23, 39
Brocken bow 135, 261
Brocken Spectre 135, 136, 261
broken corona 133, 261
Burney's halo 170, 261

cameras 198, **238**
capillary wave 94, 261
cardinal point 261
castles in the air 58, 261
cat's eyes 7, 261
cat's paws **96**, 261
caustic 93, 261
caustic network 93, 261
celestial horizon 34, 261
Cellini's halo 128, 261
chain lightning 151, 261
channel lightning 261
chromatic adaptation 233, 261
chromosphere 203, 261
circumhorizontal arc **176**, 261
circumpolar stars 261
circumscribed halo 181, 261
circumzenithal arc 165, **176**, 261
cirrocumulus 253, 261
cirrostratus 261
cirrus 253, 254, 261
civil twilight 33, 261
cloud banding 135
cloud bow 127, 261
cloud contrast bow **127**, 244, 261
clouds **139**, 247
  blocking **144**

color **143**
  dark 140, 142
  drop sizes 140
  lenticular 135, 248, 252
  mamma 254
  orographic 248, 249, 252
  pile-of-plates 247, 248
  pyrocumulus 251
  rocket-launch 257
  rotor 249
  silver lining **147**
  skirts 247
  strobing 153
  white **140**
cloud-to-stratosphere lightning 151, 261
color 21, 31, 72, 73, 233, 240
  after-images **234**
  blue moon **149**
  clouds **143**
  complementary 233
  eye (vision) 230, 233
  frost 167
  green flash 49
  halos 264
  lightning 151
  oil-on-water 97
  purple light **43**
  shadows 1, 4
  snow 159, 161
  stars 221
  sun 22
color constancy 233, 261
comets **218**
  Halley's 219
  Ikeya–Seki 218
  Oort cloud 218
  sungrazer 218
composite flash 261
condensation trail 261

cones 230, 261
conjunction 261
contrail 5, **150**, 261
contrast 262
contrast triangle 262
cornfield effect 6, 262
corona (pollen) 129
corona (solar) 262
corona (water drop) 129, 262
corona discharge 151, 262
coronae **129**, 262
corposant 262
counter-glow 262
counter-sun 262
counter-twilight 262
crepuscular arch 38, 262
crepuscular rays 16, 262
cross 187, 262
crown flash 151, 262
cumulonimbus 262
cumulus 256, 262
cyanometry 262

dark segment 38, 262
dawn 33, 43, 262
daylight 21, 262
dendrite 262
dew 109, 262
dewbow 111, 262
dewdrops 109, 128
diamond dust 262
Diamond Ring 202, 262
differential refraction 262
diffraction 243, 262
diffuse aurora 262
diffuse reflection 262
dip 262
dispersion 117, 161, 262
dissipation trail 150, 262
distrail **150**, 262
double rainbow 109, 262
drapery 262
drizzle 262
dusk 33, 262
Dutheil's halo 170, 262

earthlight 214, 262
earth shadow 34, **38**, 262
earthshine 204, **214**, 262
eclipse (moon) 214, 263
    dates 215
eclipse (sun) 263
    annular 205
    danger 204
    dates 207
    partial 205
    phenomena **204**
El Chichon 44
electromagnetic spectrum 263
electrometers 263
elves 153, 263
entropic halo 263
epoch 263
equatorial coordinate system 263
equinox 263
evening star 263
extended source 263
extinction 263
eye **230**
    after-images **234**
    blind spot 230
    chromatic adaptation 233
    color constancy 233
    color vision 230, **233**
    floaters **237**
    focal length 230
    irradiation **234**
    Mach bands **234**
    optical elements 230
    photopic vision **230**
    pupil size 230
    Purkinje effect 230, 233, 234
    resolution **233**
    response time **233**
    scotopic vision **232**
    sensitivity 230, 232

falling star 217, 263
fallstreak hole 150, 263
fallstreaks 263
false dawn 220, 263

Fata Bromosa 58, 263
Fata Morgana 58, 61, 263
ferruginous water 98, 263
Feuille's halo 170, 263
fireball 151, 263
firnspiegel 263
flaming aurora 263
'flattened sun (moon)' 47, 263
floaters **237**, 263
fluorescence 263
fog **148**, 257, 263
    ice fog 178
    radiation fog 148
    visibility in 148
    water content 148
fogbow **125**, 263
forked lightning 151, 263
forward scattering 33, 263
fovea 230, 263
Fraunhofer line 263
frost 159, 263

Galle's arc 263
galaxies **224**
gegenschein **220**, 263
geographic horizon 47, 263
geometric horizon 47, 263
geostationary satellite 208, 263
'Gibson girl' 212, 264
glaciation 140, 264
glare ice 264
glaze 264
glitter 83, 264
    cat's paws **96**
    dependence on water roughness 87
    path 83, 92
    sun on horizon 85, 92
globe lightning 151, 264
gloom 264
glory **135**, 264
godalli 150
golden bridge 83, 264
gravity wave 264
great circle 264

green flash 49, 264
    green rim 49, 51, 264
    red rim 51
    violet flash 50
gutation 109, 264

Hafgerdingar effect 58, 264
Haidinger's brush 236, **264**
Hall's halo 170, 264
halo of 22° **166**, 264
halo of 46° 169, 264
halo (ice crystal) 159, 264
    small ring (22°) **166**
    large ring (46°) **169**
    arcs 170
    catalogs **188**
    circular 180
    diffraction 167, **187**
    displays **187**
    elliptical **187**
    on snow (frost) 167
    Parry arcs 184
    pillars 165, **178**
    polarization 163, 167
    subsuns 183
    sundogs (parhelia) **171**
halo (water drop) 137
    Cellini's 128
'hare in moon' 212, 264
harvest moon 211, 264
Hasting's anthelic arc 188
haze **150**, 264
heat lightning 264
heiligenschein **128**, 264
heliac arcs **185**
Hevel's halo 171, 264
highway mirage 55, 264
Hillingar effect 58, 264
Hissink's halo 187
hoar frost 160
'holes in cloud' 264
horizons **47**, 264
    astronomical 47
    elevation dependence 47
    geometric 47

sea-level 47
sky brightness 24
skylight shift 92
horizontal rainbow 264
hunter's moon 264
hydrometer 264
hyperopic vision 264

ice **160**, 264
ice blink 144, 264
Ignis Fatuus 264
illusions **235**
    contrast triangle 262
    mach effect 29
    moon 235
    searchlight 235
index of refraction (ice) 161, 162
index of refraction (water) 117
induced color 233, 264
inferior mirage 55, 264
infralateral arcs 182, 264
interference 98, 99, 264
intracloud lightning 151, 264
iridescence 133, 265
irisation **133**, 251, 257, 258, 265
    lenticular clouds 135
irradiation **234**, 265

Jacob's ladder 265

Kage Fuji 265
Krakatoa 44
Kern's arc 176, 190, 265

lacunosus 253, 265
landpools **89**, 265
land sky 144, 265
lateral arcs **186**, 265
lateral mirage 62, 265
laurence 265
Lenard effect 151, 265
lenticular cloud 135, 248, 265
light 230, 265
    airlight **29**
    from night sky 213, **224**

from stars 226
from water **71**
purple light 34
sunlight 21, 31
twilight **33**
light chaos 265
'light of the night sky' 224, 265
lightning **151**, 265
    ball 151
    bead 151
    color 151, 154
    continuing current 151
    forked 151, 152
    heat 151
    Lenard effect 265
    ribbon 151
    St. Elmo's fire 151, 268
    sheet 269
    time resolved photography 151, **155**
Liljequist 187
limb darkening 198, 265
line lightning 265
lithometer 265
local horizon 47, 265
Local Standard Time 240, 265
looming 55, 265
low sun **31**
    brightness 31
    color 31
    flattening **47**, 54
    distortions **47**, **54**
    green flash **49**
    horizon over water 49
Lowitz arcs 182, **186**, 265
luminescence 265
luminous night cloud 192, 265
lunar eclipse 214, 265
lunar rainbow 111, 231, 265
lunar twilight 265

Mach effect 29
magnitude 221, 265
man in the moon 212, 265
marine rainbow 117, 265
mean solar time 265

mesolateral arc 182, 265
mesosphere 265
mesospheric clouds 266
meteors **217**
    bolide 217
    fireball 217
    grazing **218**
    meteorite 266
    meteoroid 266
    showers **217**, 266
    sporadic 217
midnight sun 266
Mie scattering 266
Milky Way 220, **224**
minimum deviation angle 116, 266
Minnaert, M. Preface
mirages **55**, 266
    desert 55
    highway 55
    inferior **55**
    lateral **62**
    mock 52
    superior **58**, 60
mirror image 77, 266
mist 266
mist bow 111, 266
mock sun 52, 266
moon **211**
    albedo 211
    brightness with phase 213
    color of 211
    distortion of 48
    eclipse of 214, 215
    hare in moon 212
    illusion 236
    lady in moon 212
    librations 212
    new moon, timing of 212
    phases **212**, 213
moon bow 111, 266
moon circles **88**, 266
moon dog 171
moonlight 266
morning star 266
mother-of-pearl clouds 193, 266

mountain shadows 12, 266
mountain shadow spike 14, 266
'moving ripples in clouds' 266
Mt Pinatubo 44
multiple scattering 26, 266
multiple suns (moons) 266
myopic vision 266

nacreous clouds **193**, 266
nautical twilight 34, 266
neutral points 27, 266
nightglow 266
noctilucent clouds **192**, 266
nocturnal twilight 69, 266
non-polar aurorae 266
northern lights 63, 266
Novaya Zemlya 58, 266

oblique anthelic arcs 267
occultation 267
okta 267
Olbers' paradox **226**, 267
opacity 24
opposition effect 6, 267
optical depth 24, 25, 267
optical manhole 79, 267
out-of-focus viewing 110, **240**

parafocal vision 267
paranthelic arcs **175**, 267
paranthelion 175, 267
paraselene 171, 267
parhelic circle **180**, 267
parhelion 165, 171, 267
Parry arcs **184**, 267
partial eclipse 205, 267
pearl lightning 267
penumbra 1, 267
peripheral vision 232, 267
permanent aurora 63, 267
perspective **7**, 267
phosphenes 267
phosphoresence 267
phosphorescent sea **75**
photopic vision 230, 267

photosphere 198, 267
pile of plates 247, 248
pillar **178**, 183, 267
pilot's bow 109, 267
pinhole effect 237, 267
planets 216
    crescent of Venus 216
    'evening star' 216
    magnitudes 216
    retrograde motion 217
plate crystal 160, 267
point source 1, 267
polarization (eye) 236
polarization (sky) 22, **26**
polarization (water) 80
polarizers **238**
    aircraft windows 244, 245
    polaroid 238
    safety glass 239
    Savart plate 238
primary color 267
primary rainbow 116, 268
projection mirage 268
prominence 203, 268
pulsating arcs 268
Purkinje effect 230, 233, 268
purple light 34, **43**, 268
Putnin's arcs 268

quadrature 268
quantum 268

radiant 268
radiation 268
rain 268
rainband 268
rainbows **109**, 268
    Alexander's dark band 112, **122**
    broken bow 117, 118
    double 109
    flattened drops 120, 121
    fogbow **125**
    infrared bow 111
    lunar bow 111
    marine bow 117

minimum deviation angle 116
mist bow 111
    polarization **123**
    primary **116**
    reflection bow **124**
    secondary 117, **119**
    supernumerary **119**
    tertiary, etc. **122**
    ultraviolet bow 114
    wheel 115
Rankine's halo 170, 268
Rayleigh scattering 23, 268
rays **16**, 268
    anticrepuscular 16, 18
    crepuscular 16, 18
    'rays in water' 79, 268
Rays of Buddha 16, 268
red arcs 268
red rim 50, 268
red snow 162, 268
red tide 75, 268
reflection 7, 268
reflection on water 103
reflection rainbow 124, 268
refraction 268
    normal atmosphere 46
    abnormal atmosphere 54
refraction index 268
    air 46
    water 78, 117
retroreflection 7, 268
ribbon lightning 151, 268
rime 268
'road-to-happiness' 83, 268
rocket-launch clouds 257
rods 232, 268
'Ropes of Maui' 268
rosy zone 268

St Elmo's fire 151, 268
satellites, artificial **208**
    'Aries Flasher' 209
    geostationary 208
    Molniya orbit 208
    photography 210

scattering 268
    airlight **29**
    backscattering 33
    blue sky 22
    forward 33
    near horizon 24
    patterns 33
    polarization **26**
    twilight **33**
Scheiner's halo 170, 268
Schlesinger's halo 187
schlieren 53, 268
scintillation 53, 268
scotopic vision 232, 268
sea glint 268
sea-level horizon 47, 268
'searchlight effect' 268
secondary rainbow 119, 268
seeing 53, 268
self-centered rays 102, 268
shadows 1
    alpenglow (absence of) 41
    aureole effect 102
    brightness 4
    Brocken Spectre **11**
    color 4
    contrail 4
    earth **38**
    mountain **12**
    mountain shadow spike 14, 15
    opposition effect 6
    partial eclipse 205
    penumbra 1
    triangular 11, 12
    umbra 1
    water (in) 101, 102
shadow bands 53, 200, 268
shadow hiding 6, 268
shadow of earth 268
sheet lightning 268
shimmer 268
shooting star 268
sidereal time 268
silver lining 268
sinking 268

skylight 239, 268
    from water 87
    shift towards horizon 92
sky map 144, 268
skypools **89**, 268
sleet 268
slicks **97**, 268
smog **150**, 268
smoke **150**, 255
snow 268
    sparkles **159**
    glints **159**
    color in 159, **161**, 162
snow blink 146, 268
snowflake 160, 268
solar annular eclipse 205
    future eclipses 207
solar corona 201
solar partial eclipse 205
solar points 268
solar radiation 231, 268
solar total eclipse **199**
    Baily's beads 200
    chromosphere 203
    corona 201
    coronal transients 204
    danger to eyes 204
    Diamond Ring 200, 202
    future eclipses 207
    moon's shadow 202
    prominences 203
    shadow bands 200
    shadow crescents 201
    solar diameter inferred 200
    visibility of stars 203
solstice 198, 268
south dawn 268
southern lights 268
spectre bow 135, 268
Spectre of the Brocken 135
specular reflection 268
sprites 153, 268
spurious bows 268
squinting 268
stephanome 268

stars **221**
 clusters **223**
 colors 221
 constellations **223**
 double **223**
 daytime visibility **223**
 Magellanic cloud 223
 magnitude scale **221**
 novae **223**
 numbers visible 221
 starlight 226
 variable 223
stooping 55, 57, 270
stratospheric clouds 270
streaks on water 97, 270
subanthelic point 270
subsolar point 270
subsun 183, 270
sun 270 *see also* low sun 197
 angular diameter 197
 limb darkening 198
 naked eye sunspots **198**
 on horizon 51, 85
 path in sky 197
 visibility of disk through clouds **149**
sun beam 270
sun cross 187, 270
sun dog **171**, 270
sun drawing water 270
sun-glints 245
sunlight 21, 270
sun pillar **178**
sunspots **198**, 270
 cycle 198
sunstreak **178**, 270
superior mirage **58**, 270
supernumerary rainbows **119**, 270
supralateral arcs 182, 270
surf mirage 61, 270
surface tension wave 96, 270
sylvan shine 129
syzygy 270

tangent arcs **180**, 270
temperature inversion 55, 58, 270
tertiary rainbow **122**, 270
time **240**
 universal (UT) **240**
total internal reflection 72, 270
towering 55, 57, 270
triboelectric lightning 270
Tricker's arcs 186, 191
troposphere 33, 270
true horizon 47
twilight 270, **34**
 alpenglow 34
 antitwilight arch 34, **38**
 twilight arch 34, **36**, 66
 twilight wedge 34
 volcanic activity 44
twinkling **53**, 270

Ulloa's bow 270
ultraviolet light 111, 230, 232, 271
umbra 198, 271
Umov effect 271
Universal Time **240**, 271
urban glows **224**, 225, 271

Van Buijsen's halo 170, 271
violet flash 50
volcanic eruptions **44**, 151

water **71**
 aureole effect **102**
 color from particles in **73**
 color of pure **72**
 ferruginous 98
 foam **99**
 horizon visibility **80**, 92
 interference films 97
 light from **71**, 92, 245
 mirror images **77**
 oil on troubled **97**
 optical manhole **79**
 polarization 72, **80**, 245
 reflections on **103**, 245
 reflectivity 72

 refraction through **78**, 95
 shadows on **101**
 transmission 72
 waves and the horizon 85
 wave streaming 81
 wet spot **100**
waterdrop optics **137**, 138
water sky 271
Wegener's anthelic arcs 186, 271
'wet spot' **100**, 271
white 22, 271
white dew 271
white night 271
whiteout 148, 271
white rainbow 271
will o' the wisps 271

zenith distance 23, 271
zodiacal light **220**, 224, 271